SpringerBriefs in Computer Science

Series Editors
Stan Zdonik
Shashi Shekhar
Jonathan Katz
Xindong Wu
Lakhmi C. Jain
David Padua
Xuemin (Sherman) Shen
Borko Furht
V.S. Subrahmanian
Martial Hebert
Katsushi Ikeuchi
Bruno Siciliano
Sushil Jajodia
Newton Lee

SpringerBriefs present concise summaries of cutting-edge research and practical applications across a wide spectrum of fields. Featuring compact volumes of 50 to 125 pages, the series covers a range of content from professional to academic.

Typical topics might include:

- A timely report of state-of-the art analytical techniques
- A bridge between new research results, as published in journal articles, and a contextual literature review
- A snapshot of a hot or emerging topic
- An in-depth case study or clinical example
- A presentation of core concepts that students must understand in order to make independent contributions

Briefs allow authors to present their ideas and readers to absorb them with minimal time investment. Briefs will be published as part of Springer's eBook collection, with millions of users worldwide. In addition, Briefs will be available for individual print and electronic purchase. Briefs are characterized by fast, global electronic dissemination, standard publishing contracts, easy-to-use manuscript preparation and formatting guidelines, and expedited production schedules. We aim for publication 8–12 weeks after acceptance. Both solicited and unsolicited manuscripts are considered for publication in this series.

More information about this series at http://www.springer.com/series/10028

Shaowei Wang

Cognitive Radio Networks

Dynamic Resource Allocation Schemes

 Springer

Shaowei Wang
Nanjing University
Nanjing
Jiangsu
China

ISSN 2191-5768   ISSN 2191-5776 (electronic)
ISBN 978-3-319-08935-5 ISBN 978-3-319-08936-2 (eBook)
DOI 10.1007/978-3-319-08936-2
Springer Cham Heidelberg New York Dordrecht London

Library of Congress Control Number: 2014947334

Printed on acid-free paper

Springer is part of Springer Science+Business Media (www.springer.com)

To my parents, wife, daughters and sister, for always being there for me.

Shaowei Wang

Preface

The current static spectrum management policies have caused a severe dilemma of spectrum scarcity versus under-utilization. Accordingly, the emerging Cognitive Radio (CR) technology, which allows dynamic spectrum access, has been widely deemed as a promising candidate to improve the utilization of the precious natural resource-radio spectrum for the next generation of wireless communication. Given the space-time-frequency variation in wireless communication environment, dynamic resource allocation severs as a pivotal issue to achieve the high performance of CR systems.

This brief is intended to provide a summary survey of different dynamic resource allocation schemes in CR systems, with focuses on, particularly, the spectral-efficiency and energy-efficiency in wireless communication networks. With a summary introduction of the landscape of CR technology, we detail the dynamic resource allocation problem for its motivation and challenges in CR systems. In terms of preferences, the network operator may be inclined more toward Spectral-Efficiency (SE) or Energy-Efficiency (EE) metrics in practical systems. Accordingly, the Spectral- and Energy-Efficient resource allocation schemes are comprehensively investigated in this monograph, respectively. Besides, through extensive analysis, we also explore the interrelationship between the SE and the EE, and further throw new insights into the SE-EE trade-off for operating strategies. We hope that this brief will be used as a reference for practicing engineer and researchers in the field of wireless communications.

Nanjing 210023, China   Shaowei Wang

Preface

Contents

Chapter 1
Introduction

Built on a software-defined radio, cognitive radio (CR) [1, 2] is generally defined as an intelligent wireless communication paradigm with the awareness of its environment, which is able to learn from the environment and adapt to statistical variations in the input stimuli using understanding-by-building methodology. It is proposed to achieve efficient radio spectrum utilization, as well as high reliable communication whenever and wherever needed.

1.1 Cognitive Radio Technology

CR technology allows secondary users (SUs), also referred to as CR users, to sense the radio environment and access the radio spectrum licensed by primary users (PUs), as long as the SUs do not worsen the performance of the PUs. In order to meet the requirements of opportunistic access, the physical layer of a CR system should be very flexible, which necessitates multicarrier methods to operate in CR networks. Orthogonal frequency division multiplexing (OFDM), which offers a high flexibility in radio resource management, is deemed as an appropriate air interface of a CR system [3, 4].

1.1.1 Cognitive Radio Characteristics

With the staggering increase of new wireless applications, the last decade has witnessed the looming exhaustion of radio spectrum. On the other hand, the traditional spectrum management policy, which generally assigns the radio spectrum in a fixed manner, becomes a bottleneck for high efficient spectrum utilization. As disclosed by the report of Federal Communication Commission (FCC), the temporal and geographic variations in the utilization of the licensed spectrum range from 15 to 85 % [5], as illustrated in Fig. 1.1. Consequently, such dilemma starves for innovative

© The Author(s) 2014
S. Wang, *Cognitive Radio Networks*, SpringerBriefs in Computer Science,
DOI 10.1007/978-3-319-08936-2_1

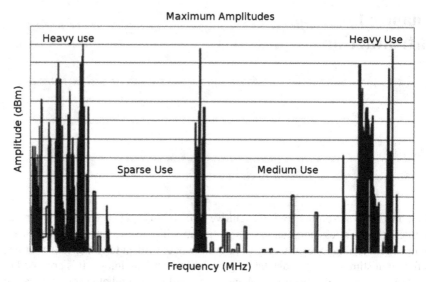

Fig. 1.1 Spectrum usage [7]

technologies to improve the usage efficiency of radio spectrum [6]. CR technology is proposed as a potential paradigm to address the crisis of spectrum scarcity.

As a vital technology for future communications, CRs are equipped with cognitive capability and reconfigurability [7, 8], enabling more efficient and flexible utilization of the limited radio resource. Specifically, cognitive capability refers to the ability of learning from the environment to obtain necessary information, such as transmitted waveform, radio frequency spectrum, communication network type, geographical information and user requirements. After gathering their desired information, the radios/devices can dynamically adapt their transmission parameters, such as transmission power, frequency, modulation type, etc., to the sensed environment variations and achieve optimal performance, known as reconfigurability.

1.1.2   Cognitive Radio Functions

Generally, the operation of a CR network consists of three fundamental tasks, including (1) radio-scene analysis, (2) channel identification and (3) power control and spectrum management [8, 9]. Through interaction with radio environment, these three tasks constitute a basic cognitive cycle, as illustrated in Fig. 1.2.

To be specific, the process of radio-scene analysis includes the estimation of interference temperature of the radio environment or the detection of spectrum holes. Generally, a band of frequencies assigned to a primary user, but not being used by that user at a particular time or geographic location is termed a spectrum hole, as shown in Fig. 1.3. The task of channel identification encompasses the estimation of

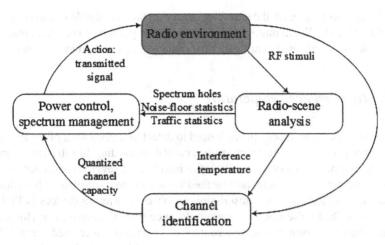

Fig. 1.2 Basic cognitive cycle [8]

channel state information and the prediction of channel capacity for utilization by the transmitter. In dynamic spectrum management, an SU may share the spectrum with PUs, other SUs, or both, while the spectrum rights are owned by primary systems. Thus, the high spectrum efficiency is mainly rooted in an appropriate spectrum sharing mechanism between the primary and the secondary networks. When SUs coexist

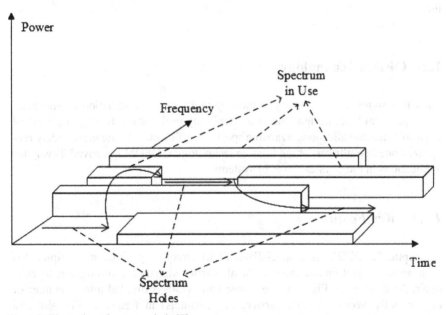

Fig. 1.3 Illustration of spectrum hole [7]

with PUs in a licensed band, the interference level to the PUs should be limited under a certain threshold. When multiple SUs share a same portion of spectrum, their access should be coordinated to mitigate their mutual collisions and interference.

1.1.3 Interference Temperature

In secondary spectrum access, the SUs need to detect the appearance of the PUs and decide which portion of the spectrum is available according to different metrics. To protect the performance of the PUs, the interference level accumulated on the licensed spectrum that is being used by the PUs should be controlled within a tolerable range. Traditionally, the transmission power of interfering devices is limited under a prescribed noise floor at a certain distance from the transmitter. However, the unpredictable appearance of new sources of interference increased the mobility and variability of radio frequency emitters, making this approach problematic. To overcome this difficulty, the FCC spectrum Policy Task Force has recommended a more practical metric for interference assessment, the *interference temperature* [10], to describe the interference limit at the receiver side. In this recommendation, the *interference temperature* is introduced as a measurement to qualify and manage the sources of interference in a radio environment, reflecting the power generated by other noise sources [11]. Accordingly, the *interference temperature* limit provides a maximum amount of tolerable interference for a given frequency band at a particular location. That is to say, any unlicensed transmission in this band must keep their interference to a licensed receiver below the interference temperature limit.

1.2 OFDM Technology

For CR systems, it is necessary to employ multi-carrier modulation technologies in the physical layer design, due to its flexibility in dynamically adapting spectral environments and allocating available spectrum efficiently. As the most widely recognized one in multi-carrier modulation technologies, OFDM has paved its way for applications in the physical layer of systems.

1.2.1 Key Features

Conceptually, OFDM is a specialized FDM (frequency division multiplexing) with an additional characteristic that all carrier signals are orthogonal to each other. As descript in Fig. 1.4, the whole bandwidth is divided into a number of orthogonally overlapping subcarriers (subchannels) in frequency domain, and the transmission on each subchannel can adaptively adopt different modulation

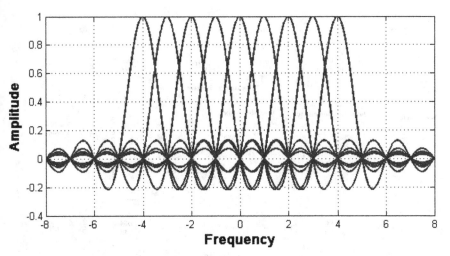

Fig. 1.4 OFDM subcarriers

schemes according to its channel condition. In this case, the data-bearing symbol stream is split into several low-rate streams to be transmitted over different sub-channels. Thus, this splitting reduces the data rate and increases the symbol duration, making it possible to combat the time dispersion effect of multipath channels in wireless environment. Moreover, the inter-symbol interference (ISI) can be further removed by extending the OFDM symbol with a cyclic prefix. Besides, the primary advantages of OFDM also include simplified channel equalizers, high spectrum efficiency, robustness against narrow-band co-channel interference, efficient implementation using Fast Fourier Transform (FFT), low sensitivity to time synchronization errors, etc.

1.2.2 OFDM-Based CR Systems

Why OFDM is a good fit for CR systems? In general, the underlying sensing and spectrum shaping capacities of OFDM, together with its flexibility in allocating resources, to fulfill the requirements of CR, make it an ideal air interface for CR systems. Here, the strength of OFDM in CR systems is elaborated from five aspects.

- Spectrum sensing. The inherent FFT cores in OFDM enables easy sensing and selection of spectrum holes without extra hardware or computation [12, 13].
- Spectrum shaping. The unique signaling of OFDM provides flexibility of being shaped adaptively to the required spectrum mask by disabling a set of subchannels.
- Adaptation to the environment. Due to its inherent features, an OFDM-based system is able to adaptively change the modulation scheme, coding and transmission

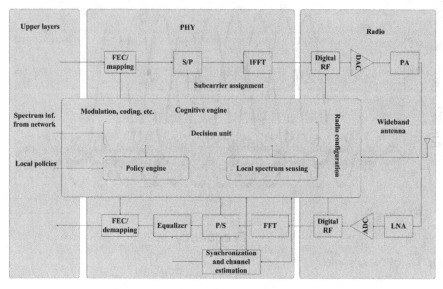

Fig. 1.5 OFDM-based CR system block diagram [4]

power of each subchannel [14]. Also, it can change the subcarrier spacing and the length of cyclic prefix according to the transmission environment and system requirements [15].

- Multiple access and spectral allocation. OFDMA (orthogonal frequency division multiple access) facilitates flexible multiple accessing and spectral allocation for CR without extra hardware complexity, where subchannels are grouped into several sets to be allocated to different users.
- Interoperability [16]. OFDM is the best candidate to achieve the ability of two or more systems to exchange information and to use the information that was exchanged, for instance, the exchange of information between the primary and secondary system in CR networks. It has been successfully used in various technologies including IEEE 802.11a and IEEE 802.11 g wireless local area network (WLAN) standards and WiMAX [17].

In this brief, we assume a CR system, also referred to as a secondary system, operates in the licensed band using OFDM. A typical block diagram of the OFDM-based CR system is illustrated in Fig. 1.5.

1.2.3 Summary

As an arising technology, there are three major features defining a CR, that are, the ability to sense, learn and adapt. These characteristics make the CR unique from other spectrum sharing and wireless communication techniques. This chapter provides a summary introduction of CR concept, characteristics and tasks. Furthermore, according to the requirements of CR systems, we also introduce a promising paradigm for the physical layer of a CR system—OFDM technology. The unique features of OFDM technology are also detailed to manifest its superiority for application in CR systems.

References

1. J. Mitola *et al.*, "Cognitive radio: Making software radios more personal," *IEEE Pers. Commun.*, vol. 6, no. 4, pp. 13–18, Aug. 1999.
2. J. Mitola, "Cognitive radio: An integrated agent architecture for software defined radio," Dissertation, Doctor of Technology, Royal Inst. Technol. (KTH), Stockholm, Sweden, 2000.
3. B. Farhang-Boroujeny and R. Kempter, "Multicarrier communication techniques for spectrum sensing and communication in cognitive radios," *IEEE Commun. Mag.*, vol. 46, no. 4, pp. 80–85, April 2008.
4. H. Mahmoud, T. Yucek and H. Arslan, "OFDM for cognitive radio: merits and challenges," *IEEE Wireless Commun.*, vol. 16, no. 2, pp. 6–15, Apr. 2009.
5. Federal Communications Commission, " Spectrum Policy Task Force," *ET Docket*, no. 02-135, Nov. 2002.
6. Federal Communications Commission, "Facilitating opportunities for flexible, efficient and reliable spectrum use employing cognitive radio technologies: Notice of proposed rulemaking and order," *ET Docket*, no. 03-108, Dec. 2003.
7. I. F. Akyildiz, W.-Y. Lee, M. C. Vuran, and S. Mohanty, "Next generation/dynamic spectrum access/cognitive radio wireless networks: A survey," *Comput. Netw.*, vol. 50, pp. 2127–2159, May 2006.
8. Simon Haykin, "Cognitive radio: brain-empowered wireless communications," *IEEE J. Sel. Areas Commun.*, vol. 23, no. 2, pp. 201–220, Feb. 2005.
9. B. Wang, K. J. R. Liu, "Advances in cognitive radio networks: A survey," *IEEE J. Sel. Topics Signal Process.*, vol. 5, no. 1, pp. 5–23, Feb. 2011.
10. Federal Communications Commission, "Establishment of interference temperature metric to quantify and manage interference and to expand available unlicensed operation in certain fixed mobile and satellite frequency bands," *ET Docket*, no. 03-289, 2003.
11. P. J. Kolodzy, "Interference temperature: A metric for dynamic spectrum utilization," *Int. J. Netw. Manage.*, vol. 16, no. 2, pp. 103–113, Mar. 2006.
12. M. Wylie-Green, "Dynamic Spectrum Sensing by Multiband OFDM Radio for Interference Mitigation," *IEEE DySPAN*, pp. 619–25, 2005.
13. T. Weiss, J. Hillenbrand, and F. Jondral, "A Diversity Approach for the Detection of Idle Spectral Resources in Spectrum Pooling Systems," *Proc. 48th Int'l. Scientific Colloq.*, Ilmenau, Germany, Sep. 2003.
14. T. Keller and L. Hanzo, "Adaptive Modulation Techniques for Duplex OFDM Transmission," *IEEE Trans. Vehic. Tech.*, vol. 49, no. 5, pp. 1893–1906, Sep. 2000.
15. D. T. Harvatin and R. E. Ziemer, "Orthogonal Frequency Division Multiplexing Performance in Delay and Doppler Spread Channels," *Proc. IEEE VTC*, vol. 3, May 1997.
16. IEEE Standard Computer Dictionary: *A Compilation of IEEE Standard Computer Glossaries*, IEEE Comp. Soc. Press, 1990.
17. "IEEE standard for local and metropolitan area networks part 16 and amendment 2," *IEEE Tech. rep. 802. 16e*, Feb. 2006.

Chapter 2
Dynamic Resource Allocation

Dynamic Resource Allocation is an essential technique to exploit the time-space-frequency variation in wireless channels by adaptively distributing precious radio resources, such as spectrum and power, to either maximize or minimize the concerned network performance metrics. In traditional static resource allocation strategies, subchannels are distributed in a predetermined manner; that is, each user is assigned fixed frequency bands regardless of the channel status. In this case, the resource allocation problem reduces to power allocation or bits loading on each subchannel, which fails to fully exploit the potential of multiuser diversity in wireless environment.

2.1 Resource Allocation in OFDM Systems

In multiuser OFDM systems, a typical resource allocation scheme might be designed to determine the subchannels, the transmission power and the bits allocated to users to optimize the desired performance.

2.1.1 Wireless Channel Characteristics

In wireless signal transmission, multipath reflections from different objects give rise to a common phenomenon-multipath fading. Specifically, the multipath reflections result in the electromagnetic wave travelling along different paths with varying length, and the interaction between those waves causes multipath fading with frequency selectivity.

Another characteristic of the multipath channel is its time-varying nature. This time variation occurs owing to the mobility of either the transmitter or the receiver, and therefore the location of reflectors in the transmission path, which gives rise to multipath, will change over time. Consequently, wireless channel is generally assumed to be wideband time-varying frequency-selective multipath fading.

© The Author(s) 2014

S. Wang, *Cognitive Radio Networks*, SpringerBriefs in Computer Science,
DOI 10.1007/978-3-319-08936-2_2

The coherence bandwidth, defined as a range of frequencies over which the channel can be considered as flat fading, is a significant parameter to characterize the multipath fading channels in frequency domain [1]. By taking advantage of this property, each OFDM subchannel can be assumed to undergo flat fading as long as the bandwidth of each subchannel is chosen much smaller than the coherence bandwidth of the channel. One of the most popular models to illustrate the statistical nature of flat fading channels is the Clark's model based on scattering [1]. In this model, the fading parameter of the channel is considered to be a random variable with Rayleigh distribution. Besides, it is also assumed that additive white Gaussian noise (AWGN) imposes on all subchannels of all users.

Accordingly, the fading parameters for different users are mutually independent, which implies that it is unlikely for a certain subchannel to undergo deep fading for all users; that is, each subchannel is expected to be in good condition for some users in the system, also referred to as multiuser diversity. One of the crucial principles of dynamic resource allocation is intended to assign each subchannel to the user with the best channel gain on it. The essential premise of such a scheme is the perfect estimation of channel information and feedback to base stations, while the channel information is always available at the beginning of each transmission block. In addition, the fading rate of each channel is supposed to be slow enough that the time-varying channel can be deemed quasi-static where the channel condition does not change within each OFDM transmission block.

The resource allocation schemes discussed in this brief are based on the assumption of perfect channel information at both the transmitter and the receiver sides. While it is rarely possible for the transmitter to obtain perfect channel state information (CSI), typical channel estimation is accurate enough to justify the use of adaptive resource allocation. The channel prediction algorithms have been studied extensively in the past years and recently, more studies have been focused on channel prediction methods in the context of OFDM [2–8].

In a multiuser OFDM system, the subchannel allocation is generally carried out based on the instantaneous channel information and the desired requirements of system. According to the aforementioned analysis about wireless channel characteristics, Figs. 2.1 and 2.2 depict the channel gain of a frequency selective fading with 32 subchannels and the channel gain of a wireless channel with eight subchannels and four users, respectively. These two figures manifest two important properties of channel gains in a multiuser OFDM system: First, different subchannels of a certain user experience different fading levels because of frequency selectivity of a channel, namely, frequency diversity; second, the subchannel gains of users at different locations vary independently, known as multiuser diversity. Taking full advantage of the channel information and its properties, the transmitter performs dynamic subchannel and power allocation to achieve the best performance of the system [9].

2.1.2 FFT-Based Transceiver

Consider the downlink of a multiuser OFDM system, a base station communicates with multiple users with limited resources, such as the given bandwidth and the

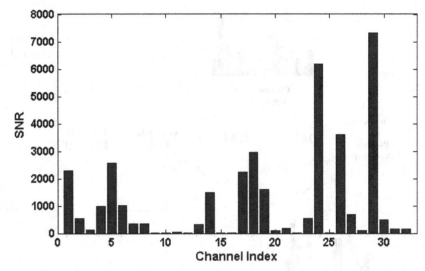

Fig. 2.1 Frequency selective channel with 32 subchannels

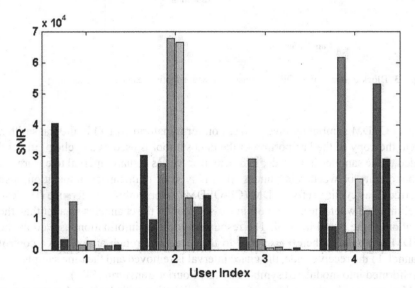

Fig. 2.2 Wireless channel with eight subchannels and four users

transmit power. The framework in a multiuser OFDM transmitter with adaptive resource allocation is illustrated in Fig. 2.3.

The process can be detailed as follows. The transmitter utilizes channel information to perform dynamic resource allocation algorithm, working out the subchannels allocated to each user, the number of bits transmitted and power loaded on each subchannel. At the output of the modulators, the complex symbols are transformed

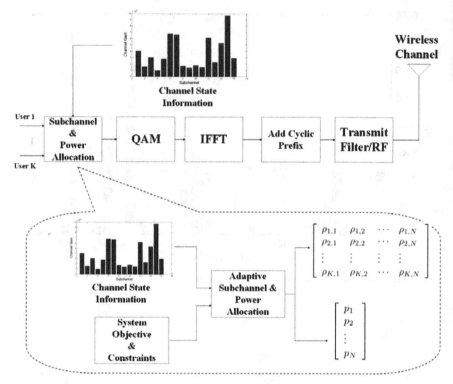

Fig. 2.3 Block diagram of an OFDM transmitter with adaptive resource allocation [9]

into an OFDM symbol by inverse fast Fourier transform (IFFT) in the transmitter. Then, the copy of the last portion of the data symbol is used as a cyclic prefix and added to the samples in time domain, which serves as a guard interval to ensure the orthogonality between subchannels. In this type of multicarrier modulation, also referred to as cyclic prefix OFDM (CP-OFDM), intersymbol interference (ISI) can be eliminated when the amount of time dispersion of the channel is smaller than the duration of the guard interval. The resource allocation information, as well as the OFDM symbols, is then transmitted to the receiver through an exclusive control channel. In the receiver side, the guard interval is removed and the time samples are transformed into modulated symbols by fast Fourier transform (FFT).

Mathematically, the problem of resource allocation with N OFDM subchannels and K users is to determine the subchannel allocation indices $\rho_{k,n}$'s, describing whether the subchannel n should be assigned to user k, and the amount of power should be loaded on subchannel n used by user k, denoted by $p_{k,n}$. The detailed block in Fig. 2.1 gives a distinct overview of this problem.

In practical systems, different users are generally not allowed to share the same subchannel to simplify the transceiver implementation; that is, the subchannel allocation indices $\rho_{k,n}$'s are mutually exclusive with binary value $\{0, 1\}$. Theoretically, it has been proved that the overall throughput can be maximized when each subchannel is allocated to only one user who enjoys the best channel gain for that subchannel, along with the optimal power loading on each subchannel [10]. Thus, throughout this brief, we assume that each subchannel can only be occupied by one user.

As the dramatic evolvement of various wireless services, future wireless communication networks are expected be flexible enough to handle myriad quality requirements, e.g., high data rate, fairness and low latency. Therefore, different dynamic resource allocation schemes and optimization techniques will be introduced in this brief in the light of different desired objectives and constraints.

2.1.3 Efficiency and Fairness

Efficiency and fairness are always two pivotal issues for dynamic resource allocation. Given limited resources for a wireless communication system, the concept of efficiency specifies either spectral efficiency (SE) or energy efficiency (EE). Specifically, SE is generally defined as the data rate per unit bandwidth, which can be calculated by dividing the total throughout by the total bandwidth of a system. Note that it considers the total data rate of the system rather than the individual achieved rate of users. Consequently, in order to achieve high SE, users with poor channel conditions might suffer relatively low data rate. On the other hand, EE is defined as the system throughput per unit energy consumption, referred to as "*bits-per-Joule*", which is mostly considered during network operation for resource allocation [11–18]. Most previous research efforts put little emphasis on EE for wireless network design until very recently. It, however, has become an inexorable trend as the blossom of green communication. Unfortunately, the SE and the EE are not always consistent and even conflict with each other sometimes. Hence, how to balance the two metrics deserves deep study to establish a more flexible and intelligent wireless networks.

In terms of fairness, the definition varies with different design criteria, indicating how equally the resources are allocated among users. For instance, it can be defined either with respect to bandwidth with the same number of subchannels assigned to each user [19], or with respect to power where each user is allocated equal portion of power from the total budget [20], or with respect to data rate which tries to achieve the same data rate of each user [21]. Particularly, when the objective is to guarantee the rate proportionality among users, it is called optimization with constrained-fairness. Generally, fairness index for rate proportional constraints is introduced as in to flexibly describe the instantaneous fairness among users. When the fairness index reaches the maximum value of 1, it indicates the fairest case in which all users would achieve the same data rate. An alternative criterion to guarantee absolute fairness among users is the max-min strategy; that is, to maximize the minimal data rate of users ensures that each user shares the equal data rate eventually.

Often, it fails to optimize the efficiency and the fairness simultaneously, since different metrics often lead to different, or even opposite design criteria for network operation. For example, optimal SE may scarify the fairness among users, because users with good channel condition are more preferable to be allocated with more resources to achieve high system throughput. Also, the optimal SE and EE can hardly be achieved at the same time due to the conflict between these two metrics in some cases. Thus, it is of great importance to figure out the trade-offs between these metrics, concerning the sorted requirement of practical systems.

2.1.4   Classes of Dynamic Resource Allocation

As one of the most important issues in OFDM systems, adaptive resource allocation has attracted significant attentions during the past two decades, particularly from the perspective of the SE. A comprehensive survey can be found in and references therein.

Resource allocation in conventional OFDM systems can be classified into two categories, namely margin adaptive [22–24] and rate adaptive [10, 21, 25–32]. The optimization objective of the margin adaptive is generally to minimize the consumed power under given rate requirements of users, while the algorithms for the rate adaptive usually try to maximize the throughput of OFDM systems under transmission power limitation. Particularly, with a given bandwidth, the rate adaptive is equivalent to the SE maximization.

Recently, the increasing awareness of green communication, which emphasizes on the EE, provides a new sight and inspiration for future wireless system design, promoting new waves of research and standard development activities. Compared with a large amount of work has been devoted to enhancing the throughput or the SE of OFDM systems, the EE is not much concerned previously. However, the energy-efficient resource allocation has been put on the agenda in both industry and academia, especially for the OFDM-based system, which serves as the most promising modulation technique for future wireless networks. Thus, an emerging class of resource allocation is catalyzed by the development of green communication, known as energy-efficient resource allocation [13–17, 33–35], which aims to maximize the overall EE of OFDM systems.

2.1.5   General Problem of RA in Multiuser OFDM Systems

Without loss of generality, we consider a multiuser OFDM system with K users denoted by the set $K = \{1, 2, \ldots, K\}$. The total bandwidth W is divided into N subchannels, denoted by $N = \{1, 2, \ldots, N\}$. The data rate of the kth user, R_k, in bits/s is given by

$$R_k = \frac{W}{N} \sum_{n=1}^{N} \rho_{k,n} \log_2 \left(1 + \frac{\gamma_{k,n}}{\Gamma} \right), \qquad (2.1)$$

where $\rho_{k,n}$ is the subchannel allocation index indicating whether the kth user occupies the nth subchannel. If subchannel n is assigned to user k, $\rho_{k,n} = 1$; otherwise, $\rho_{k,n} = 0$. $\gamma_{k,n}$ represents the signal-to-noise ratio (SNR) of the nth subchannel used by the kth user, which is given by

$$\gamma_{k,n} = p_{k,n} H_{k,n} = \frac{p_{k,n} h_{k,n}^2}{N_0 W / N}, \qquad (2.2)$$

where $p_{k,n}$ and $H_{k,n}$ denote the amount of power allocated to the kth user over the nth subchannel and the channel-to-noise ratio of the nth subchannel used by the kth user, respectively. $h_{k,n}$ signifies the channel gain for the kth user on the nth subchannel, and $N_0 \frac{W}{N}$ is the noise power over each subchannel with N_0 as the power spectral density of additive white Gaussian noise (AWGN). Γ is the SNR gap in practical modulation schemes, where an effective SNR should be adjusted according to the modulation scheme for a desired bit-error-rate (BER). The power loss, defined as the difference between the SNR for achieving a certain data rate for a practical system and that for the theoretical limit, is called the SNR gap. Thus, the SNR gap can be described as a function of BER.

For instance, the BER for an AWGN channel adopting multilevel quadrature amplitude modulation (MQAM) and ideal coherence phase detection is bounded by [36]

$$BER \leq 2e^{-1.5\gamma/(M-1)}, \qquad (2.3)$$

where $M = 2^r$ with r denoting the number of bits. On condition that $r \geq 2$ and $0 \leq \gamma \leq 30dB$, BER can be approximated within 1 dB by [37]

$$BER \leq 0.2^{-1.5\gamma/(M-1)}, \qquad (2.4)$$

and the SNR gap Γ is

$$\Gamma = \frac{-In(5BER)}{1.5}. \qquad (2.5)$$

For simplicity, the subchannel and power allocation problem is written into a general form as follows,

$$\min_{\rho_{k,n},P_{k,n}} \quad f\left(\rho_{k,n},P_{k,n}\right)$$

$$s.t. \quad C1: \rho_{k,n} \in \{0,1\}, \forall k,n$$

$$C2: \sum_{k=1}^{K} \rho_{k,n} = 1 \tag{2.6}$$

$$C3: P_{k,n} \geq 0$$

$$C4: \sum_{k=1}^{K}\sum_{n=1}^{N} \leq P_t$$

$$C5: User\ Rate\ Requirements$$

where $C1$ and $C2$ indicate each subchannel can be assigned to only one SU. $C4$ gives the total power limit for the OFDM network. $C5$ implies either fixed or variable rate requirements of users. For instance, $C5$ can be fixed rate requirement of individual or the overall throughput, or it can be proportional rate constraints to preserve the fairness among users.

Concerning the three types of dynamic resource allocation introduced in the previous section, the objective for the subchannel and power allocation can be respectively described as follows,

- Margin adaptive

$$\min_{\rho_{k,n},P_{k,n}} f(\rho_{k,n},P_{k,n}) = \sum_{k=1}^{K}\sum_{n=1}^{N} \rho_{k,n}P_{k,n}.$$

- Rate adaptive (equivalent to Spectral efficient RA)

$$\max_{\rho_{k,n},P_{k,n}} R_T = \frac{W}{N}\sum_{k=1}^{K}\sum_{n=1}^{N}\rho_{k,n} \log_2\left(1+\frac{P_{k,n}H_{k,n}}{\Gamma}\right),$$

which can be equally written into

$$\min_{\rho_{k,n},P_{k,n}} f(\rho_{k,n},P_{k,n}) = -\frac{W}{N}\sum_{k=1}^{K}\sum_{n=1}^{N}\rho_{k,n} \log_2\left(1+\frac{P_{k,n}H_{k,n}}{\Gamma}\right).$$

- Energy-efficient RA

The metric of EE is defined as the system throughput per unit energy consumption,

$$\eta_{EE}(\rho_{k,n},P_{k,n}) = \frac{R_T}{P_T} = \frac{\dfrac{W}{N}\sum\limits_{k=1}^{K}\sum\limits_{n=1}^{N}\rho_{k,n}\log_2\left(1+\dfrac{P_{k,n}H_{k,n}}{\Gamma}\right)}{\sum\limits_{k=1}^{K}\sum\limits_{n=1}^{N}\rho_{k,n}P_{k,n}}.$$

Thus the objective of optimization problem is

$$\max_{\rho_{k,n}, P_{k,n}} \eta_{EE}(\rho_{k,n}, P_{k,n}),$$

or

$$\min_{\rho_{k,n}, P_{k,n}} f(\rho_{k,n}, P_{k,n}) = -\eta_{EE}(\rho_{k,n}, P_{k,n}).$$

In short, the problem in each class is formulated accordingly and the optimal solution is obtained by different optimization techniques. Sometimes, suboptimal algorithms are preferred in real-time practical applications, due to the unaffordable computational complexity for achieving optimal solutions.

2.2 Resource Allocation in CR Systems

For the arising OFDM-based CR networks, dynamic resource allocation is of paramount importance because it is the prerequisite to achieve high system performance, such as capacity and quality of service (QoS), with limited resources. Resource allocation in an OFDM-based CR network, however, is more complex than that in a conventional OFDM system since the PUs may not adopt OFDM modulation, leading to the interference between the two systems. Moreover, the unavoidable sensing errors in the CR network can aggravate the interference, and the interference introduced to the PUs must be carefully controlled below a predefined threshold to prevent the unacceptable degeneration of the performance of the PUs.

2.2.1 Primary/Secondary Network Models

From both theoretical and practical perspectives, there are two general types of CR networks. One is the infrastructure-based network where multiple SUs are served by a common access point (AP), as illustrated in Fig. 2.4a; the other is the ad-hoc manner, as shown in Fig. 2.4b, which consists of multiple distributed secondary links.

In the infrastructure CR network, the SUs communicates with the CR AP, to some extent, the CR base station; that is, the common AP is usually correspond to one particular cell in a CR cellular network, coordinating the transmission of secondary system. In this case, the downlink and uplink transmission can be modeled as a broadcast channel and a multiple access channel, respectively. With regard to ad-hoc secondary networks, because a secondary terminal can be both a transmitter and a receiver, it is generally described as an interference channel. For a certain receiver, it will receive the signals from the direct-link channel from its corresponding transmitter, as well as the signals from the cross-link channels from

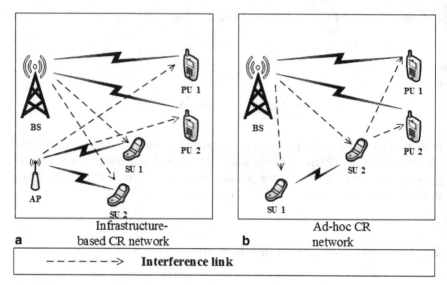

a Infrastructure-based CR network **b** Ad-hoc CR network

------ -> **Interference link**

Fig. 2.4 CR networks. **a** Infrastructure-based manner. **b** Ad-hoc manner

other transmitters. In the both types of networks, the mutual interference between primary system and secondary system should be taken into consideration.

In order to perform the dynamic resource allocation based on the CSI of SUs, the channels are assumed constant for a transitory and fixed transmission frame, such as one frequency-bin in OFDM systems. Generally speaking, the secondary transmitting terminal needs to satisfy two types of power constraints for dynamic resource allocation: CR transmission power budget and the limitation of interference level to the PUs.

2.2.2 CR Operation Models

To date, in spite of assorted operation models proposed for CR networks, there is no consensus on the terminology used for the associated definitions yet [38, 39]. By and large, there are two most popular basic operation models for CR: opportunistic spectrum access and spectrum sharing.

Specifically, the former allows SUs to transmit over the non-active frequency bands, namely spectrum holes, where no PUs are transmitting over this band. In this case, spectrum sensing serves as an indispensable technique to enable the implementation of opportunistic spectrum access; that is, the SUs try to detect the active PU transmissions over the band individually or cooperatively, and then decide to transmit on the non-active bands, where the PUs' signals are inactive with a high probability, indicated by the spectrum sensing results. Spectrum sensing is a hot issue in CR technology and interested readers may refer to [40–43] to have a close overview of the state-of-the-art development in this area.

As a counterpart, the scenario that SUs are allowed to transmit simultaneously with PUs at the same bands even if they are carrying PUs' signals is termed spectrum sharing. The permission of secondary access, however, rests on the premise that the performance degradation of each PU caused by the interference introduced by the SUs is within a tolerable range. While the heated debate on which operation model is preferable for the deployment of CR in practical systems has intensified in recent years, there is still a lack of rigorous comparative studies for these two models, concerning the tradeoffs between spectrum efficiency and implementation cost.

Since the spectrum sensing model allows simultaneous transmission of SUs and PUs in the same band, it can more efficiently capitalize on the limited spectrum compared with opportunistic spectrum access model. Therefore, in this brief, we will mainly focus on the spectrum sharing model, which is also more effective and tractable for CR to manipulate the interference to PU links.

2.2.3   Interference Management

One of the main challenges in CR networks lies in the interference management for both opportunistic spectrum access and spectrum sharing. The mutual interference between the primary and the secondary system makes the resource allocation problem in CR networks much more complex than that in traditional OFDM systems.

Although it is true that opportunistic spectrum access, which always utilizes the spectrum holes, can theoretically avoid generating interference to the active bands used by PUs, perfect spectrum sensing is too difficult to achieve in practical wireless scenarios. Even if some subchannels adjacent to active PU band are set as guard bands and not used by the SUs, it can only reduce, instead of eliminating the interference caused by imperfect spectrum sensing. There are typically two kinds of sensingerrors. The first is misdetection, which occurs when the CR system fails to detect the active PU signal over a certain subchannel and identify it as vacant. The other of sensing error is known as false alarm, which indicates that the CR system identify a subchannel is unavailable while it is vacant actually. The former will incur interference to the PUs while the latter will lower the spectrum utilization. The active PU bands, spectrum holes, guard bands and spectrum sensing errors at a particular location and time are illustrated in Fig. 2.5. For spectrum sensing model, the simultaneous transmission of the PUs and the SUs at the same band will apparently result in the co-channel interference.

To protect the PUs, we have to impose a constraint on the maximum interference power at the receiver of each PU, also referred to as interference temperature. Specifically, an interference-temperature limit provides a worst case to characterize the radio environment at a particular frequency band as well as a particular geographic location, where the receiver could be expected to operate satisfactorily. The recommendation of interference temperature contributes to an accurate measurement for

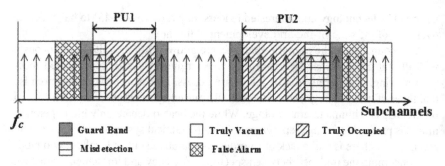

Fig. 2.5 Active bands, guard bands, spectrum holes and CR OFDM subchannels

an acceptable interference level at the receiver side in the frequency band of interest. It serves as the maximal tolerance for potential interference power that can be introduced into this band; that is, given a particular frequency band in which the interference temperature is not exceeded, it indicates that this band is available to users not serviced.

Taking advantage of interference-temperature limit, the interference management in dynamic resource allocation can be mathematically described by limiting the interference power at each PU band below a prescribed threshold.

2.2.4 General Problem of RA in OFDM-Based CR Systems

Consider the downlink of an OFDM-based CR system with K SUs, denoted by $K = \{1, 2, ..., K\}$, coexisting with L PUs in a licensed system. The CR system adopts OFDM modulation and operates in a centralized manner, that is, an AP serves all SUs in the CR network, just as a conventional base station does. Many practical wireless systems, however, employ different modulation schemes currently or in the foreseeable future. So the primary system is not assumed to use OFDM here. We assume that perfect CSI is available at the transceivers of the SUs and the PUs. The whole available bandwidth W is divided into N subchannels, denoted by $N = \{1, 2, ..., N\}$. The bandwidth of the nth subchannel spans from $f_0 + (n-1)B$ to $f_0 + nB$, where f_0 is the starting frequency and $B = W / N$.

Assume the lth PU's nominal band ranges from f_l to $f_l + B_l$, where f_l and B_l are the lth PU's starting frequency and bandwidth, respectively, the interference introduced to the lth PU by an SU' access on the nth subchannel with unit transmission power is

$$I_{n,l}^{SP} = \int_{f_l - f_0 - (n-1/2)B}^{f_l + B_l - f_0 - (n-1/2)B} g_{n,l}^{SP} \phi(f) df, \tag{2.7}$$

where $g_{n,l}^{SP}$ represents the power gain from the CR AP to the lth PU's receiver on the nth subchannel. $\phi(f)$ is the baseband power spectral density (PSD) of OFDM signal with $\phi(f) = T\left(\dfrac{\sin \pi fT}{\pi fT}\right)^2$, where T is OFDM symbol duration.

Similarly, the interference cast by the lth PU into the nth subchannel used by the kth SU is given by

$$I_{k,n,l}^{PS} = \int_{f_0 + (n-1)B - f_l - B_l/2}^{f_0 + nB - f_l - B_l/2} g_{k,n,l}^{PS} \phi_l(f) df, \qquad (2.8)$$

where $\phi_l(f)$ is the PSD of the lth PU's signal and $g_{k,n,l}^{PS}$ is the power gain from the lth PU's transmitter to the receiver of the kth SU over the nth subchannel.

Let $\gamma_{k,n}$ denote the transmission rate of the nth subchannel used by the kth SU

$$\gamma_{k,n} = \log\left(1 + \frac{P_{k,n}|h_{k,n}|^2}{\Gamma\left(N_0 B + I_{k,n}^{PS}\right)}\right), \qquad (2.9)$$

where $P_{k,n}$ is the power distributed to the nth subchannel of the kth SU and $h_{k,n}$ is the channel gain from the CR AP to the kth SU's receiver over the nth subchannel. N_0 is the PSD of noise and Γ is the SNR gap. For an uncoded MQAM, Γ is a function of a given bit-error-rate (BER) requirement with $\Gamma = -In(5BER)/1.5$. The interference $I_{k,n}^{PS}$ caused by the PUs' signals on the nth subchannel used by the kth SU can be either calculated by $\sum_{l=1}^{L} I_{k,n,l}^{PS}$ using Eq. (2.8), or regarded as noise and measured by the receiver of the kth SU [44].

The transmission power allocated to the kth SU is

$$P_k = \sum_{n=1}^{N} \rho_{k,n} P_{k,n},$$

and the achievable rate of the kth SU is

$$R_k = \frac{W}{N} \sum_{n=1}^{N} \rho_{k,n} \log_2\left(1 + \frac{P_{k,n} H_{k,n}}{\Gamma}\right).$$

The energy-efficiency in *bits/Joule* of the CR network can be accordingly obtain by

$$\eta_{EE} = \frac{\sum_{k=1}^{K} R_k}{\sum_{k=1}^{K} P_k}.$$

Generally, the optimization objective of resource allocation in CR networks basically coincide with that in OFDM systems as introduced in the previous sections, such as consumed power minimization, rate/spectral-efficiency maximization and energy-efficiency maximization. However, the additional constraint on the interference power at each PU evaluated by the interference temperature makes the problem more intricate. That is to say, the interference introduced by the secondary access to each PU must be regulated below the preset threshold I_l^{th} recommended by the interference temperature. Mathematically, these additional interference constraints should be added into the optimization problem (2.6) as concerning the RA in CR systems, which are

$$\sum_{k=1}^{K}\sum_{n=1}^{N}\rho_{k,n}P_{k,n}I_{n,l}^{SP} \leq I_l^{th}.$$

Thus, many existing RA algorithms for OFDM systems are no longer suitable.

2.3   Summary

This chapter outlines the dynamic resource allocation in both multiuser OFDM systems and OFDM-based CR networks. We have detailed the motivation and essence of dynamic resource allocation, and also provided its general mathematical form. With emphasis on system efficiency and user fairness, dynamic resource allocation is a vital technique to achieve frequency and multiuser diversity. Besides, the burgeoning CR technology evokes a new study branch for resource allocation, which integrates the interference management into the problem and formulates a more complicated optimization problem.

In a nutshell, inspired by the nature of wireless channel, dynamic resource allocation in OFDM-based systems is expected to optimize the preferred performance of system by flexibly allocating subchannels and power to users. The algorithms to solve these problems will be elaborated in depth in the following chapters, particularly focused on the spectral- and energy-efficient resource allocation in OFDM-based CR networks [45–51].

References

1. T. S. Rappaport, *Wireless Communications*. Prentice Hall PTR, 2002.
2. A. Duel-Hallen, S. Hu, and H. Hallen, "Long-range prediction of fading signals," *IEEE Signal Process. Mag.*, vol. 17, pp. 62–75, May 2000.
3. A. Forenza and R. W. Heath, "Link adaptation and channel prediction in wireless OFDM systems," *in Proc. 45th IEEEMWSCAS*, vol. 3, pp. 211–214, Aug. 2002.

4. M. Sternad and D. Aronsson, "Channel estimation and prediction for adaptive OFDM down-links [vehicular applications]," in *Proc. IEEE VTC*, vol. 2, pp. 1283–1287, Oct. 2003.
5. I. C. Wong, A. Forenza, R. W. Heath, and B. L. Evans, "Long range channel prediction for adaptive OFDM systems," in *Proc. IEEE ACSSC*, vol. 1, pp. 732–736, Nov. 2004.
6. I. C. Wong and B. L. Evans, "Joint channel estimation and prediction for OFDM systems," in *Proc. IEEE Globecom'05*, vol. 4, pp. 2255–2259, Dec. 2005.
7. D. Schafhuber and G. Matz, "MMSE and adaptive prediction of time varying channels for OFDM systems," *IEEE Trans. Wireless Commun.*, vol. 4, pp. 593–602, Mar. 2005.
8. I. C. Wong and B. L. Evans, "Low-complexity adaptive high-resolution channel prediction for OFDM systems," in *Proc. IEEE Globecom'06*, Nov. 2006.
9. S. Sadr, A. Anpalagan, and K. Raahemifar, "Radio resource allocation algorithms for the downlink of multiuser OFDM communication systems," *IEEE Commun. Surv. & Tutor.*, vol. 11, no. 3, pp. 92–106, Sep. 2009.
10. J. Jang and K. B. Lee, "Transmit power adaptation for multiuser OFDM systems," IEEE J. Select. Areas *Commun.*, vol. 21, pp. 171–178, Feb. 2003.
11. Y. Chen, S. Zhang, S. Xu, and G. Li, "Fundamental trade-offs on green wireless networks," *IEEE Commun. Mag.*, vol. 49, no. 6, pp. 30–37, June 2011.
12. D. Feng, C. Jiang, G. Lim, L. Cimini, Jr., G. Feng, and G. Li, "A survey of energy-efficient wireless communications," *IEEE Commun. Surv. & Tutor.*, vol. PP, no. 99, pp. 1–12, 2012.
13. G. Miao, N. Himayat, G. Li, and S. Talwar, "Low-complexity energy efficient scheduling for uplink OFDMA," *IEEE Trans. Commun.*, vol. 60, no. 1, pp. 112–120, Jan. 2012.
14. C. Xiong, G. Li, S. Zhang, Y. Chen, and S. Xu, "Energy- and spectral- efficiency trade off in downlink OFDMA networks," *IEEE Trans. Wireless Commun.*, vol. 10, no. 11, pp. 3874–3886, Nov. 2011.
15. G. Miao, N. Himayat, G. Li, and S. Talwar, "Distributed interference aware energy-efficient power optimization," *IEEE Trans. Wireless Commun.*, vol. 10, no. 4, pp. 1323–1333, Apr. 2011.
16. D. Ng, E. Lo, and R. Schober, "Energy-efficient resource allocation in OFDMA systems with large numbers of base station antennas," *IEEE Trans. Wireless Commun.*, vol. 11, no. 9, pp. 3292–3304, Sep. 2012.
17. D. W. K. Ng, E. S. Lo, and R. Schober, "Energy-efficient resource allocation in multi-cell OFDMA systems with limited backhaul capacity," *IEEE Trans. Wireless Commun.*, vol. 11, no. 10, pp. 3618–3631, Oct. 2012.
18. Y. Pei, Y.-C. Liang, K. C. Teh, and K. H. Li, "Energy-efficient design of sequential channel sensing in cognitive radio networks: Optimal sensing strategy, power allocation, and sensing order," *IEEE J. Sel. Areas Commun.*, vol. 29, no. 8, pp. 1648–1659, Sep. 2011.
19. Y. Otani, S. Ohno, K. Ann Donny Teo, and T. Hinamoto, "Subcarrier allocation for multi-user OFDM system," in *Proc. Asia-Pasific Conf. on Commun.*, pp. 1073–1077, 2005.
20. W. Rhee and J. M. Cioffi, "Increase in capacity of multiuser OFDM system using dynamic subchannel allocation," in *Proc. IEEE VTC'00*, vol. 2, pp. 1085–1089, May 2000.
21. Z. Shen, J. G. Andrews, and B. L. Evans, "Adaptive resource allocation in multiuser OFDM systems with proportional rate constraints," *IEEE Trans. Wireless Commun.*, vol. 4, pp. 2726–2737, Nov. 2005.
22. C. Y. Wong, R. S. Cheng, K. B. Letaief, and R. D. Murch, "Multiuser OFDM with adaptive subcarrier, bit and power allocation," *IEEE J. Select. Areas Commun.*, vol. 17, pp. 1747–1758, Oct. 1999.
23. G. Zhang, "Subcarrier and bit allocation for real-time services in multiuser OFDM systems," in *Proc. IEEE ICC'04*, vol. 5, pp. 2985–2989, June 2004.
24. L. Xiaowen and Z. Jinkang, "An adaptive subcarrier allocation algorithm for multiuser OFDM system," in *Proc. IEEE VTC'03*, vol. 3, pp. 1502–1506, Oct. 2003.
25. G. Song and Y. G. Li, "Cross-layer optimization for OFDM wireless networks-Part I: theoretical framework," *IEEE Trans. Wireless Commun.*, vol. 4, pp. 614–624, Mar. 2005.

26. G. Song and Y. G. Li, "Cross-layer optimization for OFDM wireless networks-Part II: Algorithm development," *IEEE Trans. Wireless Commun.*, vol. 4, pp. 625–634, Mar. 2005.

27. Z. Shen, J. G. Andrews, and B. L. Evans, "Optimal power allocation in multiuser OFDM systems," in *Proc. IEEE Globecom'03*, vol. 1, pp. 337–341, Dec. 2003.

28. I. C. Wong, Z. Shen, B. L. Evans, and J. G. Andrews, "A low complexity algorithm for proportional resource allocation in OFDMA systems," *in Proc. IEEE Workshop on Signal Processing Systems*, Oct. 2004.

29. H. Yin and H. Liu, "An efficient multiuser loading algorithm for OFDM based broad band wireless systems," in *Proc. IEEE Globecom'00*, vol. 1, pp. 103–107, Nov. 2000.

30. G. Song and Y. G. Li, "Utility-based joint physical-MAC layer optimization in OFDM," in *Proc. IEEE Globecom'02*, vol. 1, pp. 671–675, Nov. 2002.

31. G. Song and Y. G. Li, "Adaptive subcarrier and power allocation in OFDM based on maximizing utility," *in Proc. IEEE VTC*, vol. 2, pp. 905–909, Apr. 2003.

32. M. Tao, Y.-C. Liang, and F. Zhang, "Resource allocation for delay differentiated traffic in multiuser OFDM systems," *IEEE Trans. Wireless Commun.*, vol. 7, no. 6, pp. 2190–2201, June 2008.

33. C. Xiong, G. Y. Li, S. Zhang, Y. Chen, S. Xu, "Energy-Efficient Resource Allocation in OFDMA Networks," in *Proc. IEEEGlobecom'11*, Dec. 2011.

34. C. Xiong, G. Y. Li, S. Zhang, Y. Chen, S. Xu, "Energy-Efficient Resource Allocation in OFDMA Networks," *IEEE Trans. Commun.*, vol. 60, no. 12, pp. 3767–3778, Dec. 2012.

35. A. Zappone, G. Alfano, S. Buzzi, M, Meo, "Energy-efficient non-cooperative resource allocation in multi-cell OFDMA systems with multiple base station antennas," in *Proc. IEEE GreenCom'11*, pp. 82–87, Sep. 2011.

36. G. J. Foschini and J. Salz, "Digital communications over fading radio channels," *Bell Syst. Tech. J.*, pp. 429–456, Feb. 1983.

37. A. J. Goldsmith and Soon-Ghee Chua, "Variable-rate variable-power MQAM for fading channels," *IEEE Trans. Commun.*, vol. 45, pp. 1218–1230, Oct. 1997.

38. Q. Zhao and B. M. Sadler, "A survey of dynamic spectrum access," *IEEE Signal Processings. Mag.*, vol. 24, no. 3, pp. 79–89, May 2007.

39. A. Goldsmith, S. A. Jafar, I. Mari'c, and S. Srinivasay, "Breaking spectrum gridlock with cognitive radios: An information theoretic perspective," *Proc. IEEE*, vol. 97, no. 5, pp. 894–914, May 2009.

40. Z. Quan, S. Cui, and A. Sayed, "Optimal linear cooperation for spectrum sensing in cognitive radio networks," *IEEE J. Select. Topics Signal Process.*, vol. 2, no. 1, pp. 28–40, Feb. 2008.

41. Y.-C. Liang, Y. Zeng, E. C. Y. Peh, and A. T. Hoang, "Sensing-throughput tradeoff for cognitive radio networks," *IEEE Trans. Wireless Commun.*, vol. 7, no. 4, pp. 1326–1337, Apr. 2008.

42. B. H. Juang, Y. Li, and J. Ma, "Signal processing in cognitive radio," *Proc. IEEE*, vol. 97, no. 5, pp. 805–823, May 2009.

43. Y. H. Zeng, Y.-C. Liang, A. T. Hoang, and R. Zhang, "A review on spectrum sensing for cognitive radio: challenges and solutions," *EURASIP J. Advances Signal Process.*, 2010.

44. P. Setoodeh and S. Haykin, "Robust transmit power control for cognitive radio," *Proc. of the IEEE*, vol. 97, no. 5, pp. 915–939, May 2009.

45. S. Wang, "Efficient resource allocation algorithm for cognitive OFDM systems," *IEEE Commun. Lett.*, vol. 14, no. 8, pp. 725–27, Aug. 2010.

46. M. Ge and S. Wang, "Fast optimal resource allocation is possible for multiuser OFDM-based cognitive radio networks with heterogeneous services," *IEEE Trans. Wireless Commun.*, vol. 11, no. 4, pp. 1500–1509, Apr. 2012.

47. S. Wang, Z.-H. Zhou, M. Ge and C. Wang, "Resource allocation for heterogeneous cognitive radio networks with imperfect spectrum sensing," *IEEE J. Sel. Areas Commun.*, vol. 31, no. 3, pp. 464–475, 2013.

48. S. Wang, M. Ge and W. Zhao, "Energy-Efficient Resource Allocation for OFDM-based Cognitive Radio Networks," *IEEE Trans. Commun.*, vol. 61, no. 8, pp. 3181–3191, Aug. 2013.

49. S. Wang, M. Ge, C. Wang, "Efficient Resource Allocation for Cognitive Radio Networks with Cooperative Relays," *IEEE J. Sel. Areas Commun.*, vol. 31, no. 11, pp. 2432–2441, Nov. 2013.

50. S. Wang, Z.-H. Zhou, M. Ge and C. Wang, "Resource Allocation for Heterogeneous Multiuser OFDM-based Cognitive Radio Networks with Imperfect Spectrum Sensing," In *Proc. IEEE INFOCOM'12*, pp. 2264–2272, Mar. 2012.

51. S. Wang, F. Huang and Z.-H. Zhou, "Fast Power Allocation Algorithm for Cognitive Radio Networks," *IEEE Commun. Lett.*, vol. 15, no. 8, pp. 845–847, Aug. 2011.

Chapter 3
Spectral-Efficient Resource Allocation in CR Systems

Spectral efficiency (SE), defined as the system throughput per unit of bandwidth, is a prevalent criterion for wireless network optimization. For instance, the peak value of SE is always taken as one of the pivotal indicators of 3rd Generation Partnership Project (3GPP) evolution for network performance. Particularly, in OFDM-based systems with a given bandwidth, dynamic resource allocation is always carried out to explore the potential system throughput by making efficient use of channel characteristics.

3.1 Single-User CR Systems

3.1.1 System Model

Basically, we consider the downlink of a single-user CR system coexisting and sharing radio spectrum with a licensed primary system with L PUs. The CR system is assumed to operate with OFDM modulation while it is not necessary for the PUs to adopt OFDM. The PUs are served by a base station and the SUs access the CR network via a CR AP as shown in Fig. 3.1.

The total bandwidth W is divided into N OFDM subchannels in the CR network. The bandwidth of the nth subchannel spans from $f_0 + (n-1)\dfrac{W}{N}$ to $f_0 + n\dfrac{W}{N}$, where f_0 is the starting frequency. The spectrum of the lth PU spans from f_l to $f_l + W_l$.

The interference introduced to the lth PU by the transmission on the nth subchannel with unit power is $I_{n,l}^{SP}$, given by

$$I_{n,l}^{SP} = \int_{f_l - f_0 - (n-1/2)W/N}^{f_l + W_l - f_0 - (n-1/2)W/N} g_{n,l}^{SP}\phi(f)df,$$

where $g_{n,l}^{SP}$ is the power gain from the CR AP to the lth PU's receiver. The interference to the nth subchannel cast by the transmission of the PUs is denoted by I_n^{PS},

© The Author(s) 2014
S. Wang, *Cognitive Radio Networks*, SpringerBriefs in Computer Science,
DOI 10.1007/978-3-319-08936-2_3

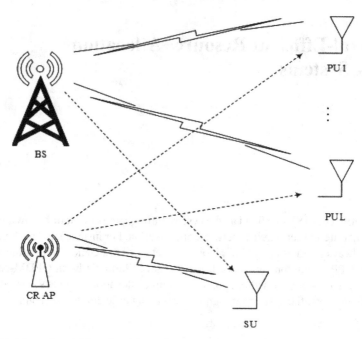

Fig. 3.1 Channel model for primary/secondary network with single SU

which can be regarded as noise and measured by the receiver of an SU. Thus, the
SNR of the nth subchannel with unit power is

$$H_n = \frac{|h_n|^2}{N_0 \dfrac{W}{N} + I_n^{PS}},$$

where N_0 is the PSD of the additive white Gaussian noise and h_n is the channel gain
from the CR AP to the receiver of the SU.

The achievable rate of the CR system can be calculated by

$$R = \sum_{n=1}^{N} \frac{W}{N} \log_2\left(1 + \frac{p_n H_n}{\Gamma}\right),$$

where Γ represents the SNR gap and we can take $\Gamma = -\dfrac{\ln(5BER)}{1.5}$ for an uncoded
MQAM with a specified BER.

3.1.2 Problem Formulation

We expect to maximize the SE of the CR network, while the interference introduced to each PU band should be controlled below a given threshold prescribed by the interference temperature. Besides, the transmission power of the CR AP is limited by a certain budget. In a signal-user CR system, all subchannels are used by the single SU and the resource allocation problem is simplified into power loading; that is, under the interference and power budget constraints, the limited power should be dynamically allocated to each subchannel in order to maximize the SE of the CR system. Thus, the concerned problem can be mathematically written into

$$\max_{p_n} \quad \sum_{n=1}^{N} \frac{1}{N} \log_2 \left(1 + \frac{p_n H_n}{\Gamma} \right)$$

$$s.t. \quad C1: \sum_{n=1}^{N} p_n \leq P_t$$

$$C2: \quad p_n \geq 0, n = 1,\ldots,N$$

$$C3: \sum_{n=1}^{N} p_n I_{n,l}^{SP} \leq I_l^{th}, l = 1,\ldots,L. \tag{3.1}$$

Here, p_n is the power allocated over the nth subchannel that needs to be optimized. The constraint $C1$ describes the total transmission limit at the CR AP, bounded by the maximum transmission power P_t. $C3$ constrains the interference power at the lth PU band below its interference tolerance I_l^{th}.

3.1.3 Fast Barrier Algorithm

According to [1], (3.1) defines a convex optimization problem with $N + L + 1$ inequality constraints. The optimization variables p_n's can be denoted by a vector p with $p = \{p_1,\ldots,p_N\}$. Generally, the global optimal solution of the problem (3.1) can be obtained by standard convex optimization techniques with a typical complexity of $O(N^3)$, such as the barrier method. However, a practical wireless system cannot afford such high computational cost for online process, since there are always thousands of subchannels in an OFDM system.

Thus, we hope to reduce the complexity while figuring out the global optimal solution of the problem. Inspired by the special structure of (3.1), we propose a fast barrier method to solve it with a much lower complexity of $O(L^2 N)$ [2]. Obviously, for a given number of PUs, the complexity of our proposal is approximately linear to the number of subchannels, dramatically reducing the high computational load in the standard algorithm.

With the barrier method, the original problem is converted into a sequence of unconstrained minimization problem by exploiting a logarithmic barrier function

Table 3.1 Barrier method

Initialization	**Given a strictly feasible starting point** p, $t = t_0$, $\mu > 1$, tolerance ϵ.
Repeat	a) *Centering Step.* Solve (3.2), starting at P, to obtain its optimal solution p^*;
	b) *Update.* $p = p^*$;
	c) *Stopping Criterion.* $(N + L + 1)/t < \epsilon$;
	d) *Increase.* $t = \mu t$.

with parameter t, which indicates the accuracy of the approximation. Then, these unconstrained optimization problems are solved by Newton method.

The barrier function of problem (3.1) is

$$\varphi(p) = -\sum_{n=1}^{N} \log(p_n) - \log\left(P_t - \sum_{n=1}^{N} p_n\right) - \sum_{l=1}^{L} \log\left(I_l^{th} - \sum_{n=1}^{N} p_n I_{n,l}^{SP}\right).$$

Let $f(p)$ denote the objective function of problem (3.1) with

$$f(p) = \sum_{n=1}^{N} \frac{1}{N} \log\left(1 + \frac{p_n H_n}{\Gamma}\right).$$

For convenient calculation in the following steps, the binary logarithmic function $\log_2(.)$ in the objective function in replaced by a natural logarithmic function $\log(.)$, which will not change the solution to the original problem.

Accordingly, the optimal solution of (3.1) can be approximated by solving the following unconstrained minimization problem [62]

$$\min_p \psi_t(p) = -tf(p) + \varphi(p) \tag{3.2}$$

with $t > 0$. The solution to each minimization problem is called a central point in the central path related to the original problem. As t increases, the central point will be more and more accurately approximated to the optimal solution of the original problem. The outline of the barrier method is detailed in Table 3.1.

The unconstrained problem (3.2) can be addressed by Newton method. Specifically, Newton method is carried out during the centering step for searching the optimal p^* to minimize $\psi_t(p)$. The procedure of Newton method is summarized in Table 3.2.

The Newton Step is given by the following equation,

$$\nabla^2 \psi_t(p)\Delta p = -\psi_t(p) \tag{3.3}$$

Table 3.2 Newton method

Initialization	A feasible starting point p, $\alpha \in (0, 1/2), \beta \in (0, 1)$, tolerance ϵ_n.
Repeat	a) Compute Newton step Δp and $\lambda = -\nabla \psi_t(p) \Delta p$;
	b) Quit if $\lambda^2/2 < \epsilon_n$;
	c) Backtracking line search on $\nabla \psi_t(p)$. $s = 1$. **while** $\psi_t(p + s\Delta p) > \psi_t(p) - \alpha s \lambda^2$ $s = \beta s$ **endwhile**
	d) Update $p = p + s\Delta p$

Where $\nabla^2 \psi_t(p)$ and $\nabla \psi_t(p)$ are the Hessian and the gradient of $\psi_t(p)$, respectively. Generally, it would cost $O(N^3)$ if the Newton step in (3.3) is worked out by matrix inversion, which introduces the major computational load of the barrier method.

In other words, if we can reduce the complexity for computing Newton step, the major limitation imposed on the barrier method for practical application will be eliminated. Based on such an expectation, we further explore the structure of the equation (3.3) and find its special feature which makes fast calculation of Newton step possible. The matrix $\nabla^2 \psi_t(p)$ is given by

$$\nabla^2 \psi_t(p) = D + \frac{\nabla f_0(p) \nabla f_0(p)^T}{f_0(p)^2} + \sum_{l=1}^{L} \frac{\nabla f_l(p) \nabla f_l(p)^T}{f_l(p)^2}$$

where $D = \text{diag}(D_1, D_2, ..., D_N)$ with $D_n = \dfrac{t}{N} \dfrac{H_n^2}{(\Gamma + H_n p_n)^2}$, $f_0 = P_t - \sum_{n=1}^{N} p_n$ and $f_l = I_l^{th} - \sum_{n=1}^{N} I_{n,l}^{SP} p_n$, $l = 1, ..., L$. For simplicity, the Hession of $\psi_t(p)$ can be consequently decomposed into a diagonal matrix and several rank-one matrices as follows,

$$\nabla^2 \psi_t(p) = D + \sum_{m=1}^{M} g_m(p) g_m(p)^T \qquad (3.4)$$

where $M = L + 1$, $g_1 = \nabla f_0(p)/f_0(p)$ and $g_{m+1} = \nabla f_m(p)/f_m(p), m = 1, ..., L$. It can be proved that the Hessian is positive definite because the diagonal matrix D has $D_n > 0, \forall n$ and all $g_m(p) g_m(p)^T > 0, m = 1, ..., M$.

Consider the special structure shown in (3.4), we derive a fast algorithm to speed up the computation of Newton step based on matrix inversion lemma [3]. Denote $g_0 = -\nabla \psi_t(p)$, the procedure of an M-step iterative algorithm is detailed in Table 3.3.

Table 3.3 Algorithm for fast computation of Newton step

Initialization	$Dv_i^0 = g_{i-1}, i = 1, \dots, M+1$
Step 1	Solve the M equations $(D + g_M g_M^T)v_i^1 = g_{i-1}$ using $$v_i^1 = v_i^0 - \frac{g_M^T v_i^0}{1 + g_M^T v_{M+1}^0} v_{M+1}^0, i = 1, \dots, M$$
Step 2	Solve the M -1 equations $(D + \sum_{j=M-1}^{M} g_j g_j^T)v_i^2 = g_{i-1}$ using $$v_i^2 = v_i^1 - \frac{g_M^T v_i^1}{1 + g_M^T v_{M+1}^1} v_{M+1}^1, i = 1, \dots, M-1$$
Step m	Solve the $M - m + 1$ equations $(D + \sum_{j=M}^{M+1-m} g_j g_j^T)v_i^m = g_{i-1}$ using $$v_i^m = v_i^{m-1} - \frac{g_{M-m+1}^T v_i^{m-1}}{1 + g_{M-m+1}^T v_{M-m+2}^{m-1}} v_{M-m+2}^{m-1}, i = 1, \dots, M-m+1$$
Step M	Solve the equation $(D + \sum_{j=1}^{M} g_j g_j^T)v_1^M = g_0$ using $$v_1^M = v_1^{M-1} - \frac{g_M^T v_1^{M-1}}{1 + g_M^T v_2^{M-1}} v_2^{M-1}, \text{ where we have } \Delta p = v_1^M$$

The iterative algorithm for the fast computation of Newton step can be explained as follows: First, we solve $M+1$ matrix systems $Dv_i^0 = g_{i-1}, i = 1, \dots, M+1$ for initialization; then we employ matrix inversion lemma to work out the one-rank update of $p_{s,n}$, v_i^1 in $(D + g_M g_M^T)v_i^1 = g_{i-1}, i = 1, \dots, M$ using the known variables v_i^0's directly. Such process is executed step by step until we solve the matrix system (3.3), which requires M step for updating.

Particularly, since D is a diagonal matrix, the $M+1$ matrix system in the initialization step can be quickly solved by

$$v_i^0 = \begin{bmatrix} D_1^{-1} & & \\ & \ddots & \\ & & D_N^{-1} \end{bmatrix} g_{i-1}, i = 1, \dots, M+1,$$

which costs $O(N)$ for solving each equation.

In terms of computational load, the complexity of the barrier method is mainly caused by the calculation of Newton step. Instead of using matrix inversion, a fast algorithm is proposed to speed up the computation of Newton step. The complexity of the fast algorithm can be analyzed as follows. The $M+1$ matrix systems for the initialization can be solved with complexity of $O(MN)$ as discussed above; then the M-step one-rank update is carried out as shown in Table 3.3. The complexity is roughly $O(M^2N)$. Thus, we can conclude that the complexity to work out the optimal solution can be measured by $O(M^2 N)$. Compared to the standard matrix inversion which generates a complexity of $O(M^2 N)$, the computational cost of our proposal is significantly reduced since it usually follows that $M \ll N$ in practical systems.

3.1.4 Numerical Results

A series of numerical experiments are conducted to evaluate the performance of our proposed algorithm. Consider an OFDM-based CR system, each PU occupies random bandwidth which spans some continuous subchannels. The noise power is 10^{-13} W and the interference threshold of all PUs are set to 5×10^{-13} W. The channel suffers from frequency selective fading. The path loss exponent is 4, the variance of logarithmic normal shadow fading is 10 dB and the amplitude of multipath fading is Rayleigh. The parameters of the barrier method are set to typical values as in and initialized with a strictly feasible solution generated by

$$p_n = \min\{P_t/N, \min_n\{\min_l\{I_l^{th}/I_{n,l}^{SP}\}\}\}, \forall n.$$

First we explore the achievable capacity of the CR system of our proposed power allocation algorithm for different numbers of subchannels with $P_t = 1$ W, $L = 2$. Here, we compare our proposed optimal power allocation scheme with other two heuristic algorithms:

1. Greedy Max-Min algorithm in [4], which is developed to solve the formulated multidimensional 0-1 knapsack problem for power allocation.
2. Efficient bit-loading algorithm in [5], which always picks the bit with lowest cost to allocated to subchannels.

As listed in Table 3.4, our proposal outperforms the other two algorithms because it can always work out the optimal solutions. On the other hand, the suboptimal solutions given by the heuristic algorithms have a significant gap with the optimal ones. From the numerical results shown in Table 3.4, it is indicated that our proposed algorithm has a visible capacity gain compared to algorithms proposed in and [5].

We also investigate the convergence performance of our proposal in terms of the number of Newton iterations in Figs. 3.2 and 3.3. As aforementioned analysis, the computational load mainly lies in the computation of Newton step. That is to say, if the number of Newton iterations is large or varies in a wide range, the algorithm would be difficult to be applied to practical wireless systems. Figures 3.2 and 3.3 show that it is not the case for our proposed method for all concerned settings. Figure 3.2 depicts the number of Newton iterations for the barrier method to converge in 100 random instances for $N = 64$ and $N = 1024$, respectively. For a given number of subchannels, the number of Newton iterations remains stable with a small average value. Compared Fig. 3.2a, b, we can find that the number of Newton iterations varies in a narrow range, despite large variation in the number of subchannels. In Fig. 3.3, we give the cumulative distribution function (CDF) of Newton iterations of the barrier method with a duality gap of less than 10^{-3} for

Table 3.4 Total number of allocated bits in the CR system with $P_t = 1$ W, $L = 2$

N (Number of subchannels)	64	128	256	512	1024
Our proposal	252	449	810	1252	1888
Algorithm in [4]	250	444	802	1238	1863
Algorithm in [5]	244	434	783	1211	1829

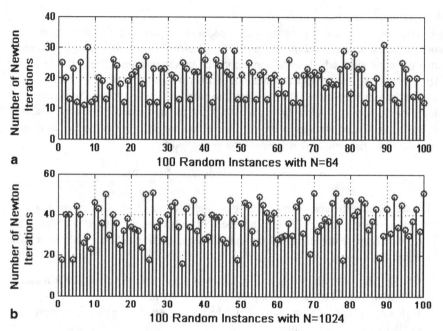

Fig. 3.2 Number of Newton iterations required for convergence for 100 channel realizations. **a** N=64. **b** N=1024

different number of subchannels. It also demonstrates the stability of the number of Newton iterations. Although it is true that more subchannels result in more Newton iterations, the increase of the number of Newton iteration is quite slow and acceptable for practical applications. All these observations validate the efficient convergence of our proposed method.

To further manifest the efficiency of our proposed algorithm, the time cost of the fast barrier method is studied as well. In Fig. 3.4, we illustrate the average elapsed time in *second* of our proposed algorithm as a function of number of subchannels over 1000 instances with $P_t = 1$ W, $L = 2$, compared to that of other heuristic algorithms in [4] and [5]. The time consumption is counted by in-built *tic-toc* function in *Matlab*. From Fig. 3.4, it is shown that the algorithm in consumes much more time than our method and also grows more sharply with the increase of the number of subchannels while it only produces suboptimal solutions. In addition, although the time cost of our method is slightly higher than that of the algorithm developed in [5], our proposal

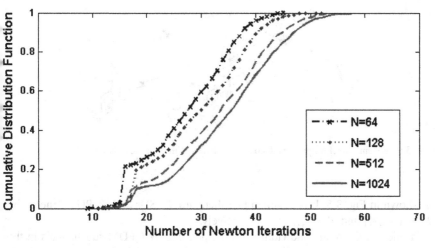

Fig. 3.3 CDF of number of Newton iterations over instances with $P_t = 1\,\mathrm{W}, L = 2$

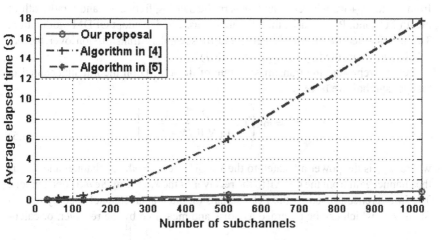

Fig. 3.4 Average elapsed time versus number of subchannels with $P_t = 1\,\mathrm{W}, L = 2$

can achieve a larger sum capacity gain. On the other hand, we can conservatively conclude that the consumed time could be further reduced for specialized computing platform, which makes our proposal promising for possible applications.

3.2 Multi-User CR Systems

3.2.1 System Model & Problem Formulation

In this section, we consider the downlink of a multiuser CR system served by an exclusive CR AP, coexisting with a licensed primary system with multiple PUs,

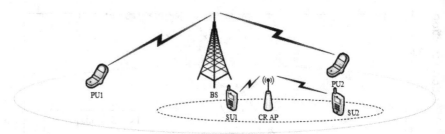

Fig. 3.5 System model for primary/secondary network with multiple SUs

as shown in Fig. 3.5. It is assumed that there are K SUs and L PUs, denoted by $\mathcal{K} = \{1, \ldots, K\}$ and $\mathcal{L} = \{1, \ldots, L\}$, respectively.

Typically, the total bandwidth W is divided into N OFDM subchannels in the CR network and the bandwidth of the nth subchannel spans from $f_0 + (n-1)W/N$ to $f_0 + nW/N$, where f_0 is the starting frequency. The spectrum of the lth PU spans from f_l to $f_l + W_l$, where f_l and W_l are the starting frequency and bandwidth of the lth PU band. Particularly, the PUs are not assumed to adopt OFDM modulation. The interference to the lth PU generated by secondary transmission over the nth subchannel is denoted by $I_{n,l}^{SP}$, given by (2.7).

Let $r_{k,n}$ represent the transmission rate of the kth SU on the nth subchannel in bits per symbol, we have

$$r_{k,n} = \log\left(1 + \frac{p_{k,n}\left|h_{k,n}\right|}{\Gamma(N_0 W/N + I_{k,n}^{PS})}\right) \tag{3.5}$$

where $p_{k,n}$ is the power allocated to the kth SU over the nth subchannel and $h_{k,n}$ is the channel gain from the CR AP to the receiver of the kth SU of the nth subchannels. The interference caused by the PU's signals on the nth subchannel used by the kth SU is $I_{k,n}^{PS}$, which can be regarded as noise and measured by the receiver, or calculated by (2.8). For notation brevity, denote $H_{k,n} = \dfrac{\left|h_{k,n}\right|^2}{N_0 W/N + I_{k,n}^{PS}}$ as the SNR of the nth subchannel used by the kth SU with unit power. Accordingly, the achievable rate of the kth SU is

$$R_k = \sum_{n=1}^{N} \rho_{k,n} \log\left(1 + \frac{p_{k,n} H_{k,n}}{\Gamma}\right) \tag{3.6}$$

where $\rho_{k,n}$'s are the subchannel assignment indices, informing whether the nth subchannel is allocated to the kth SU; that is, $\rho_{k,n} = 1$ indicates that the nth subchannel is assigned to the kth SU while $\rho_{k,n} = 0$ implies the opposite condition. In practical system, each subchannel can only be used by one SU.

To reflect the practical situations in wireless communication, heterogeneous services, for both real-time (RT) SUs and non-real-time (NRT) SUs are considered for the resource allocation optimization. Thus, the K SUs are classified into two categories: K_1 RT SUs with individual minimal rate requirement $R_{k,min}$ and $K - K_1$ NRT users. In terms of spectral efficient resource allocation, we equivalently try to maximize the sum rate of the NRT SUs while guaranteeing the minimal required rate of the RT SUs under the total transmission budget and interference limits. Mathematically, the concerned optimization problem can be described as follows,

$$\max_{p_n} \quad \sum_{k=K_1+1}^{K} \sum_{n=1}^{N} \rho_{k,n} \log\left(1 + \frac{p_{k,n} H_{k,n}}{\Gamma}\right)$$

$$s.t. \quad C1: \quad \sum_{k=1}^{K} \sum_{n=1}^{N} \rho_{k,n} p_n \le P_t$$

$$C2: \quad p_{k,n} \ge 0, \forall k, n$$

$$C3: \quad R_k \ge R_{k,min}, k = 1, \ldots, K_1$$

$$C4: \quad \sum_{n=1}^{N} \rho_{k,n} p_n I_{n,l}^{SP} \le I_l^{th}, l = 1, \ldots, L \tag{3.7}$$

$$C5: \quad \sum_{k=1}^{K} \rho_{k,n} = 1, \forall n$$

$$C6: \quad \rho_{k,n} \in \{0,1\}, \forall k, n,$$

where P_t is the transmission power limit of the CR AP, and I_l^{th} is the maximum interference tolerance of the lth PU band. In the above problem, $C1$ and $C2$ are the transmission power constraints, while C4 is the interference limits set by PUs. C3 describe the minimal rate requirements of RT SUs. C5 and C6 declare that each subchannel is kept from being shared among multiple SUs.

3.2.2   Relaxation Method

Obviously, (3.7) defines a mixed integer programming problem since both binary variables $\rho_{k,n}$'s and real variables $p_{k,n}$'s are involved. The main difficulty of solving (3.7) mainly lies in the integer constraints C6, which generates K^N possible subchannel assignments if using an exhaustive search. It is apparently impractical even for a middle number of OFDM subchannels, let alone the thousands subchannels in practical wireless systems.

An intuitive way to address such problems is the relaxation method, which relaxes the integer variables into continuous ones. To be specific, redefine $\rho_{k,n} \in [0,1]$ as the fraction of the nth subchannel allocated to the kth SU, temporarily permitting multiple SUs to share the same subchannel. Let $s_{k,n}$ characterize the actual power consumption of the kth user on the nth subchannel in a time frame interval, it follows that $s_{k,n} = \rho_{k,n} p_{k,n}$. Then, the original problem (3.7) can be converted into

$$\max_{P_n} \quad \sum_{k=K_1+1}^{K} \sum_{n=1}^{N} \rho_{k,n} \log\left(1+\frac{s_{k,n}H_{k,n}}{\Gamma\rho_{k,n}}\right)$$

$$s.t \quad C1: \quad \sum_{k=1}^{K}\sum_{n=1}^{N} S_n \le P_t$$

$$C2: \quad s_{k,n} \ge 0, \forall k,n$$

$$C3: \quad \sum_{n=1}^{N} \rho_{k,n} \log\left(1+\frac{s_{k,n}H_{k,n}}{\Gamma\rho_{k,n}}\right) \ge R_{k,min}, k=1,\ldots,K_1 \tag{3.8}$$

$$C4: \sum_{n=1}^{N} s_n I_{n,l}^{SP} \le I_l^{th}, l=1,\ldots,L$$

$$C5: \sum_{k=1}^{K} \rho_{k,n} = 1, \forall n$$

$$C6: \quad \rho_{k,n} \ge 0, \forall k,n.$$

Note that the constraint $\rho_{k,n} \le 1$ is implied by C5 and C6, which can be omitted.

It is easy to prove that (3.8) is a convex optimization problem which can be solved by convex optimization techniques. The optimal solution to the above problem serve as an upper bound of the problem (3.7) because all feasible solutions to the original problem fall into the solution space of the relaxed one. Since the solution to the above problem contains sharing factors $\rho_{k,n}$'s which are not binary variables, rounding is necessary to obtain a feasible subchannel assignment to the original problem. Then power distribution among subchannels has to be re-allocated accurately with a given subchannel assignment after rounding procedure, in order to guarantee the feasible solution to (3.7).

Generally, it requires $O((2KN+N)^3)$ for standard techniques to solve the relaxed problem (3.8). Inspired by the fast barrier method for optimal power allocation analyzed in Chap. 3, Sect. 3.1.3, can we develop a similar fast algorithm to reduce the computational cost of solving (3.8)? Based on this purpose, we further explore the structure of the above problem.

First, according to the barrier method, the optimal solution to (3.8) can be approximated by a set of equality constrained minimization problems as follows,

$$\min_{x} \quad \Psi_t(x) = -tf(x) + \phi(x)$$

$$s.t. \quad \sum_{k=1}^{K} \rho_{k,n} = 1, \forall n, \tag{3.9}$$

where $x \in \mathcal{R}^{2KN \times 1}$ is the optimization variable including $\rho_{k,n}$'s and $\rho_{k,n}$'s, i.e. $x = (S_{1,1}, \rho_{1,1}, \ldots, S_{K,N}, \rho_{K,N})$. $f(x)$ is taken as the objective function in (3.8). For simplicity, denote

$$
f_i = \begin{cases} P_t - \sum_{k=1}^{K}\sum_{n=1}^{N} s_{k,n}, & i = 1 \\[2mm] \sum_{n=1}^{N} r_{k,n} - R_{k,\min}, & k = 1,\ldots,K_1, i = k+1 \\[2mm] I_l^{th} - \sum_{k=1}^{K}\sum_{n=1}^{N} s_{k,n} I_{n,l}^{SP}, & l = 1,\ldots,L, i = K_1 + L + 1, \end{cases}
$$

we have

$$
\phi(x) = -\sum_{i=1}^{K_1+L+1} \log(f_i) - \sum_{k=1}^{K}\sum_{n=1}^{N}\left(\log(s_{k,n}) + \log(\rho_{k,n})\right).
$$

With a given parameter t during the centering step in the barrier method, Newton method is utilized to compute the central point. Newton step Δx and the associated dual variable are given by

$$
\begin{bmatrix} \nabla^2 \Psi t(x) & A \\ A^T & 0_n \end{bmatrix} \begin{bmatrix} \Delta x_{nt} \\ \nu \end{bmatrix} = \begin{bmatrix} -\nabla \psi t(x) \\ 0_v \end{bmatrix} \tag{3.10}
$$

where $0_n \in R^{N \times N}$ is a zero matrix and $0_v \in R^{N \times 1}$ is a zero vector. A is a $2KN \times N$ matrix in which $A_{2m,n} = 1, m = (k-1)N + n, \forall k,n$ and the other elements are all zeroes. For more detailed procedure of the barrier method and Newton method, refer to the Chaps. 9 and 10 in [1], similar to that in Tables 3.1 and 3.2.

We find that the Hessian of $\Psi_t(x)$ in (3.10) also can be decomposed into the similar structure as (3.4), that is

$$
\nabla^2 \Psi_t(x) = D + \sum_{i=1}^{M} F_i F_i^T \tag{3.11}
$$

where $M = K_1 + L + 1$ and $D = \mathrm{diag}(D_{1,1}, D_{1,2},\ldots,D_{k,N}) \in \mathcal{R}^{2KN \times 2KN}$ with

$$
D_{k,n} = \begin{bmatrix} \dfrac{1}{s_{k,n}^2} & 0 \\[3mm] 0 & \dfrac{1}{\rho_{k,n}^2} \end{bmatrix} + \chi_k \begin{bmatrix} \dfrac{\rho_{k,n} h_{k,n}^2}{(s_{k,n} h_{k,n} + \Gamma \rho_{k,n})^2} & -\dfrac{s_{k,n} h_{k,n}^2}{(s_{k,n} h_{k,n} + \Gamma \rho_{k,n})^2} \\[4mm] -\dfrac{sk,n h_{k,n}^2}{(s_{k,n} h_{k,n} + \Gamma \rho_{k,n})^2} & \dfrac{s_{k,n}^2 h_{k,n}^2}{\rho_{k,n}(s_{k,n} h_{k,n} + \Gamma \rho_{k,n})^2} \end{bmatrix}
$$

and $\chi_k = \begin{cases} 1/f_k & k = 1,\ldots,K_1 \\ t & k = K_1+1,\ldots,K \end{cases}$. Besides, the vectors F_i's are given by

$F_i(x) = \dfrac{\nabla f_i}{f_i}, i = 1,\ldots,M$.

Since the quasi-diagonal matrix D is positive definite and all $F_i F_i^T \geq 0$, the matrix $\nabla^2 \Psi_t(x)$ is positive definite. Moreover, because A is a full row rank matrix, the matrix in the left of (3.10) is invertible.

According to the analysis above, we find that matrix system (3.10) for Newton step also can be efficiently solved by the fast barrier method which we have proposed in Sect. 3.1.3. The idea for deriving the fast computation of Newton step is similar to that in Table 3.3, which will not be elaborated in this section. The major difference lies in the initialization step for solving $M+1$ matrix systems, which is more intricate in this problem. Thus, we intend to specify the method for efficient initialization, where the matrix systems can be written into a general form,

$$\begin{bmatrix} D & A \\ A^T & 0_n \end{bmatrix} v^0 = \begin{bmatrix} g \\ 0_v \end{bmatrix}. \tag{3.12}$$

It follows

$$\begin{bmatrix} v^0_{2i-1} \\ v^0_{2i} \end{bmatrix} = D^{-1}_{k,n} \begin{bmatrix} g_{2i-1} \\ g_{2i} - v^0_{2KN+n} \end{bmatrix}, \quad i = (k-1)N+n \tag{3.13}$$

Substituting (3.13) into (3.12), we have

$$v^0_{2KN+n} = \frac{1}{\displaystyle\sum_{k=1}^{K} D^{-1}_{k,n_{2,2}}} \sum_{k=1}^{K} (D^{-1}_{k,n_{2,1}} g_{2i-1} + D^{-1}_{k,n_{2,2}} g_{2i})$$

and then all other v^0_i's can be obtained by substituting v^0_{2KN+n} into (3.13). The computation cost is $O(KN)$. The total computational complexity of fast barrier method for solving this relaxed problem is bounded by $O(M^2 KN)$.

The solution to (3.8) is not feasible for the original problem because the $\rho_{k,n}$'s are not binary variables now. Thus, after solving the relaxed problem, a rounding procedure is carried out for subchannel assignment. Intuitively, larger $\rho_{k,n}$ implies the kth SU obtains higher rate over the nth subchannel. It is reasonable to allocate the nth subchannel to the k^*th SU who has the largest $\rho_{k,n}$ over this channel.

To ensure the feasibility of solution to the original problem, the power should be re-allocated among subchannels with the given subchannel allocation. Since the binary variables are fixed to 0 or 1, the C5 and C6 constraints in the problem (3.7) vanish. Let Ω_k denote the subchannel set of the kth SU, the optimal power allocation is formulated as follows,

$$\begin{aligned} \max_{p_{k,n}} \quad & \sum_{k=K_1+1}^{K} \sum_{n \in \Omega_k} \log(1 + p_{k,n} H_{k,n} / \Gamma) \\ s.t. \quad C_1 \quad & \sum_{n \in \Omega_k} r_{k,n} \geq R_{k,MiN}, 1, \ldots, K_1 \\ C_2 \quad & \sum_{k=1}^{K} \sum_{n \in \Omega_k} p_{k,n} \leq P_t \\ C_3 \quad & \sum_{k=1}^{K} \sum_{n \in \Omega_k} p_{k,n} I^{SP}_{n,l} \leq I^P_l, l = 1, \ldots, L \\ C_4 \quad & p_{k,n} \geq 0, \forall k, n. \end{aligned} \tag{3.14}$$

Note that the power allocation problem is also a convex optimization problem with inequality constraints, whose structure is similar to the problem (3.1). Consequently, the problem can be addressed by the proposed fast barrier method as that in Sect. 3.1.3. Similarly, the computational cost for optimal power allocation using fast barrier method is measured by $O(L^2 N)$ where L and N are the number of PUs and subchannels, respectively. More details about the fast barrier method for solving the above problem can be found in [6].

Actually, only few subchannels are shared among multiple SUs as $\rho_{k,n}$ is mostly either 1 or 0 for $K \ll N$, given by the optimal solution of the relaxed problem [7]. It means that the relaxation method can achieve approximate optimal solution to the original problem, which is also verified by the simulation results given at the end of this chapter.

3.2.3   Two-Stage Method

To further reduce the complexity, we propose a two-stage approach, which addresses the problem (3.7) by solving the subchannel assignment and power allocation separately. At the first stage, subchannel allocation is implemented via a simple but effective heuristic method, which removes the integer constraints in the original problem. During the second stage, power distribution among subchannels is carried out. It is expected to maximize the sum rate of the NRT SUs while guaranteeing the minimal rate requirements of the RT SUs under the transmission power and interference constraints.

- Subchannel assignment

The intuition of our method for subchannel assignment is as follows. In an OFDM-based CR system, an OFDM subchannel with higher SNR for an SU may also generate more interference to PUs. It indicates that interference constraints also lay an upper bound for the transmission power over each subchannel. Thus, the SNR of an OFDM subchannel and the interference introduced to the PUs should be jointly considered to measure the condition of the subchannel. Here, we propose a metric evaluate the potential capacity of a subchannel, which gives the maximum achievable rate of the nth subchannel used by the kth SU,

$$r_{k,n}^M = \log(1 + p_{k,n}^M H_{k,n}), \tag{3.15}$$

where $p_{k,n}^M$ is the maximum possible transmission power for the kth SU on the nth subchannel, which is jointly determined by the power budget and the interference to PUs,

$$p_{k,n}^M = \min\left(P_t, \min_{l \in \mathcal{L}} \left(\frac{I_l^{th}}{I_{n,l}^{SP}} \right) \right), \tag{3.16}$$

The channel capacity is normalized by (3.16), which is utilized as an assignment criterion for determining the subchannel allocation.

Table 3.5 Subchannel allocation

1.**Initialization:**
2. $\mathcal{N}_t = \mathcal{N}$, $\Omega_k = \emptyset, \forall k$
3.   Set the RT SUs' rates to zeros: $R_{t_i} = 0$, for $i = 1, 2, ..., K_1$
4.**For the RT SUs:**
5.**While** $\mathcal{N}_t \neq \emptyset$ and $R_{t_k} < R_{k,min}$ for any $1 \leq k \leq K_1$
6.   Find k^* satisfies $R_{t_{k^*}} - R_{k^*,min} \leq R_{t_k} - R_{k,min}, 1 \leq k \leq K_1$
7.   For k^*, find n^* satisfies $r^M_{k^*,n^*} \geq r^M_{k^*,n}, \forall n \in \mathcal{N}_t$
8.   Update $R_{t_{k^*}} = R_{t_{k^*}} + \log(1 + p_{k^*,n^*} H_{k^*,n^*})$
9.   Update $\Omega_{k^*} = \Omega_{k^*} \cup n^*, \mathcal{N}_t = \mathcal{N}_t \setminus n^*$
10.**Endwhile**
11.**For the NRT SUs:**
12.**For** $i = 1$ to length($\mathcal{N}_t$)
13.   For $n^* = \mathcal{N}_{t_i}$, find k^* satisfies $r^M_{k^*,n^*} \geq r^M_{k,n^*}$
14.   Update $\Omega_{k^*} = \Omega_{k^*} \cup n^*$
15.**Endfor**

Recall that Ω_k denotes the set of subchannels allocated to the kth SU, the intu-itiveness behind our subchannel allocation scheme is as follows. The RT SU whose current rate is the farthest away from his/her required rate has the priority to get a new subchannel among the available ones. Preferably, the subchannel with the high-est achievable rate associated with an RT SU will be chosen in this step. The power distributed to a subchannel is temporally set as $p_{k,n} = \min(P_t/N, \min_{l \in \mathcal{L}}(I_l^{th}/I_{n,l}^{SP}))$ to satisfy the power and interference constraints. Such procedure continues until all RT SUs meet their rate requirements. Then, each of the remaining subchannels is allocated to the NRT SU who has the highest achievable rate over this channel to roughly maximize the sum rate of the NRT SUs. The detail of subchannel allocation strategy is shown in Table 3.5.

- Power allocation

When the subchannel assignment is given, the power distribution among subchan-nels will be carried out to solve the problem (3.14). As discussed above, its optimal solution can be derived by the fast barrier method we have proposed with complex-ity of $O(L^2 N)$.

Alternatively, we also propose a heuristic power allocation algorithm to approxi-mate the optimal solution with lower complexity than the fast barrier method. In [5], the author introduces an index function to measure the cost of allocating each bit, based on which an efficient bit loading algorithm is developed. Similarly, we defined a normalized cost function indicating the cost of allocating certain rate on each subchannel,

$$F_c(r_{k,n}) = \frac{e^{r_{k,n}} - 1}{e^{r^M_{k,n}} - 1} \tag{3.17}$$

Here, the channel SNR and interference to PUs are jointly considered in the normalized cost function, reflecting the ability of carrying bits for each subchannel. Based on (3.17), the power allocation problem (3.14) can be converted into its approximated form,

$$
\begin{aligned}
\max_{r_{k,n}} \quad & \sum_{k=K_1+1}^{K} \sum_{n\in\Omega_k} r_{k,n} \\
s.t. \quad C_1 \quad & \sum_{n\in\Omega_k} r_{k,n} \geq R_{k,min}, k=1,\ldots,K_1 \\
C_2 \quad & r_{k,n} \geq 0, \forall k,n \\
C_3 \quad & \sum_{k=1}^{K} \sum_{n\in\Omega_k} F_c(r_{k,n}) \leq C.
\end{aligned}
\tag{3.18}
$$

The parameter C is a constant determined by the transmission power and the interference threshold, indicating the maximum sum capacity of all subchannels. The key idea of this transformation is to unify the constraints $C2$ and $C3$ in (3.14) into a normalized constraint $C3$ in (3.18). Note that the value of C is not necessary to be known in advance, which will be shown in the following paragraphs.

Then, KKT conditions are used to solve the problem. Without consideration of $C2$ in (3.18), the Lagrangian of the problem (3.18) is

$$
L(r_{k,n}, \lambda_0, \mu_k) = -\sum_{k=K_1+1}^{K} \sum_{n\in\Omega_k} r_{k,n} + \lambda_0 \left(\sum_{k=1}^{K} \sum_{n\in\Omega_k} F_c(r_{k,n}) - C \right)
$$
$$
+ \sum_{k=1}^{k_1} \mu_k \left(R_{k,min} - \sum_{n\in\Omega_k} r_{k,n} \right),
$$

where λ_0 and μ_k ($k=1,\ldots,K_1$) are the Lagrange multipliers with $\lambda_0 \geq 0$ and $\mu_k \geq 0$. Applying the KKT conditions, the optimal solutions $r_{k,n}^*$'s are given by the following equations when concerning $C2$ in (3.18),

$$
\frac{\partial L}{\partial r_{k,n}^*} \begin{cases} = 0, & r_{k,n}^* \geq 0 \\ > 0, & r_{k,n}^* = 0 \end{cases},
$$

$$
\lambda_0 \left(\sum_{k=1}^{K} \sum_{n\in\Omega_k} F_c(r_{k,n}^*) - C \right) = 0,
$$

$$
\mu_k \left(R_{k,min} - \sum_{n\in\Omega_k} r_{k,n}^* \right) = 0, \quad k=1,\ldots,K_1.
$$

Accordingly, it follows

$$r_{k,n}^* = \left[\log\left(\frac{\mu_k(e^{r_{k,n}^M} - 1)}{\lambda_0} \right) \right]^+, \quad k = 1, \dots, K_1 \tag{3.19}$$

For the RT SUs, while the optimal solutions for the NRT SUs are

$$r_{k,n}^* = \left[\log\left(\frac{e^{r_{k,n}^M} - 1}{\lambda_0} \right) \right]^+, \quad k = K_1 + 1, \dots, K. \tag{3.20}$$

For the kth SU, denote Ω_k^* as the subchannels set in which $r_{k,n}^* \geq 0$. Sort the $r_{k,n}^*$'s in ascending order. For $m, n \in \Omega_k^*$, we have

$$r_{k,m}^* - r_{k,n}^* = \log\left(\frac{p_{k,m}^M H_{k,m}}{p_{k,n}^M H_{k,n}} \right), \quad k = 1, \dots, K \tag{3.21}$$

Let $\omega_{m,n}^k = \frac{p_{k,m}^M H_{k,m}}{p_{k,n}^M H_{k,m}}$ and $N_k = |\Omega_k^*|$, the rate of the kth user is given by

$$R_k = N_k r_{k,1*} + \sum_{n \in \Omega_k^*} \log(\omega_{n,1*}^k), \quad k = 1, \dots, K, \tag{3.22}$$

where $r_{k,1*}^*$ denotes the rate of the minimal value among subchannels in Ω_k^*. According to (3.21) and (3.22), we can obtain the closed-form of $r_{k,n}^*$ for the RT SUs with $\mu_k \neq 0$,

$$r_{k,n} = \frac{1}{N_k}\left(R_{k,min} - \sum_{n \in \Omega_k^*} \log\left(\omega_{n,1*}^k \right) \right) + \log\left(\omega_{n,1*}^k \right).$$

For the NRT users, we have

$$\sum_{k=K_1+1}^{K} P_k = \sum_{k=K_1+1}^{K} \sum_{n \in \Omega_k} \frac{e^{r_{k,n}} - 1}{H_{k,n}} \leq P_t - P_r,$$

$$\sum_{k=K_1+1}^{K} I_k^l = \sum_{k=K_1+1}^{K} \sum_{n \in \Omega_k} \frac{(e^{r_{k,n}} - 1)}{H_{k,n}} \leq I_l^P - I_l^r, \tag{3.23}$$

where P_r and I_l^r are the power consumption and the interference generated to the lth PU generated by the RT SUs, which are determined by the allocated rate to the subchannels used by the RT SUs. The achievable rates of the NRT SUs are bounded by the remaining power budget and the tolerable interference power of each PU band. Based on (3.20) and (3.23), the minimal λ_0, which allows the maximum sum rate of the NRT SUs, can be directly solved. More details about this heuristic power allocation algorithm can be found in [6].

Particularly, our proposed two-stage method is also effective to address the resource allocation problem for multiuser CR networks with proportional rate requirements [8].

3.2.4   Numerical Results

The performance of our proposed algorithms, including relaxation method (RM) and two-stage method(TM), is evaluated through a series of experiments. For simulation, we consider a multiuser OFDM-based CR system where all users are randomly located in a 3×3 km area, and each receiver uniformly distributed in the circle within 0.5 km from its transmitter. The path loss exponent is 4. The variance of shadowing effect is 10 dB and the amplitude of multipath fading is Rayleigh. The bandwidth of each PU is assumed to be randomly generated by uniform distribution with the maximum value $2W/3L$. The noise power is set to 10^{-13} W and the interference threshold of each PU is 5×10^{-12} W.

Specifically, we proposed three different resource allocation schemes for multiuser OFDM-based CR networks, that are, RM, TM with optimal power allocation (TM-OP) and TM with approximation power allocation (TM-SOP). We intend to compare our proposals with other two algorithms: maximum SNR priority (MSP) subchannel allocation and minimum interference priority (MIP) subchannel allocation, both of which adopt optimal power allocation. For the MSP, a subchannel is always assigned to the SU with the highest SNR over this channel, while the MIP prefer allocating a subchannel to the SU whose transmission over this channel generates the minimum interference to the PUs.

Figure 3.6 illustrates the average number of bits per symbol of the NRT SUs as a function of transmission power budget. There are two PUs and four SUs, including two RT SUs with uniform rate requirements 20 bit/symbol. The total bandwidth is divided into 64 OFDM subchannels in CR system. It can be observed that the sum rate of the NRT SUs increases with the growth of power limit. The performance of our proposed algorithms is much better than the MSP-OP and the MIP-OP schemes. Particularly, our proposed RM always performs the best, while the capacity loss of the TM-SOP compared to the TM-OP is slight.

We also give the average sum rate of the NRT SUs versus the number of subchannels in Fig. 3.7. Assume $P_t = 1$ W, $K = 4$, $K_1 = 2$, $L = 2$ and $R_{k,min} = 20$ bit/symbol. The phenomenon that more subchannels contribute to more capacity can be explained by the effect of channel diversity. To put it in another way, with more available subchannels, an SU is more likely to choose subchannels with high channel gains, which will definitely result in higher system capacity. However, the MSP-OP and MIP-OP schemes, which fail to well evaluate the condition of subchannels, are unable to effectively utilize the channel diversity effect and achieve much lower

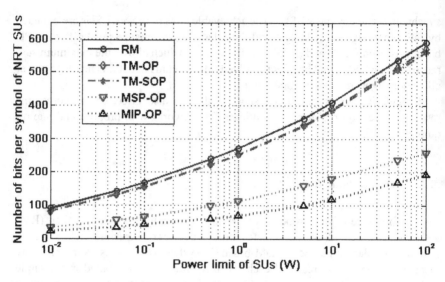

Fig. 3.6 Average number of bits per OFDM symbol of NRT SUs as a function of P_t. $R_{k,min}$ = 20 bit/symbol, $N = 64, K = 4, K_1 = 2, L = 2$

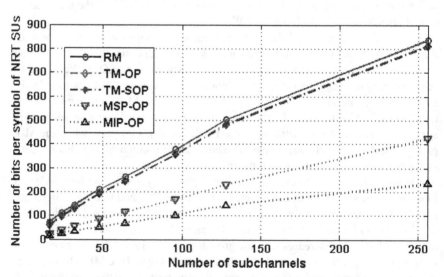

Fig. 3.7 Average number of bits per OFDM symbol of NRT SUs as a function of the number of subchannels. $R_{k,min}$ = 20 bit/symbol, $P_t = 1W, K = 4, K_1 = 2, L = 2$

capacity than our proposed algorithms. Note that when N is sufficiently large, the performance loss due to the heuristic subchannel allocation method in the TM-SOP is neglectful, which suggests its possible application in practical because there are always thousands of subchannels.

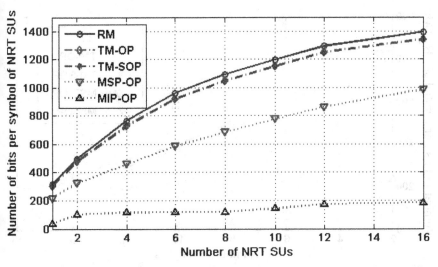

Fig. 3.8 Average number of bits per OFDM symbol of NRT SUs as a function of the number of NRT SUs. $R_{k,min} = 10$ bit/symbol, $N = 128, K_1 = 4, L = 2$

Figure 3.8 discloses another important characteristic in multiuser wireless system, that is, multiuser diversity. We have $N = 128$, $P_t = 1$ W and $R_{k,min} = 10$ bit/symbol. The number of the RT SUs is fixed to 4, while the number of the NRT SUs varies from 1 to 16. More NRT SUs lead to more chances for a subchannel to be assigned to an SU with higher channel gain, which makes more achievable capacity of the NRT SUs. The results in Fig. 3.8 are coincident with the theoretical analysis that the sum rate of the NRT users increases with more NRT SUs. Also, our proposed algorithm exhibit significant superiority over the MSP-OP and the MIP-OP.

On the other hand, we also study the sum rate of the NRT SUs versus the number of the RT SUs. Here, we set $R_{k,min} = 10$ bit/symbol, $P_t = 1W, N = 128, L = 2$ with four NRT SUs. The number of the RT SUs varies from 2 to 16. The curve shown in Fig. 3.9 reveals that the sum capacity of NRT SUs decreases when there are more RT SUs. It can be explained intuitively that more subchannels and power are consumed by the RT services when more RT SUs try to access with its minimal rate requirements. In this case, the limited radio resource becomes more frequently exhausted, which intensifies the capacity loss of the NRT users.

Synthesizing the above simulation results, the performance of our proposed algorithms is mainly evaluated in terms of capacity with various setting. In a nutshell, our proposed algorithms can always achieve more capacity than the other two schemes. The poor performance of the MSP-OP and the MIP-OP is attributed to their stereotyped subchannel allocation strategies, which fail to jointly consider the channel SNR and the interference level. For our proposed algorithms, the RM always performs the best, and the TM-SOP can achieve almost the same capacity of the TM-OP in most cases. Particularly, the remarkable performance of the RM compensates for its relatively high computational load, while both of the TM-OP and the TM-SOP can achieve about 95 % capacity of the RM.

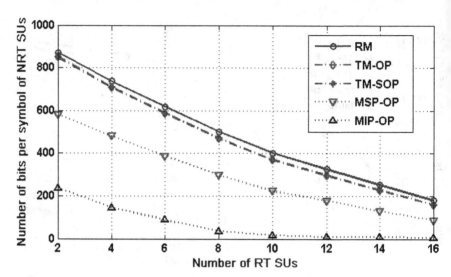

Fig. 3.9 Average number of bits per OFDM symbol of NRT SUs as a function of the number of RT SUs. $R_{k,min} = 10$ bit/symbol, $N = 128, K - K_1 = 4, L = 2$

Finally, we investigate the convergence of the fast barrier method, focusing on the number of Newton iterations. The number of Newton iterations for fast barrier method to solve the relaxed RA problem and optimal power allocation are shown in Fig. 3.10a, b, respectively. For a guaranteed duality gap of 10^{-3}, the number of Newton iterations varies in a relatively narrow range for both optimal RA based on relaxation and power allocation. Such an observation demonstrates the promising convergence performance of our proposed fast barrier method.

3.3 Relay-Enhanced CR Systems

In recent year, relay has emerged as a potential technology to utilize the spatial diversity in wireless environment. By improving the SE and enlarging the coverage area, relay communication can enhance the overall performance of wireless systems. The channel capacity for a basic three-node relay system with a source node (SN), a relay node (RN) and destination node (DN) is analyzed in [9]. There are three major relay transmission protocols: amplify-and-forward (AF), decode-and-forward (DF) and coded cooperation, as discussed in [10]. Due to its inherent advantages in wireless communications, cooperative relay has been introduced to the standard of the next generation cellular systems, such as the Long Term Evolution Advanced (LTE-Advanced).

Cooperative relay is also a good fit for CR systems. Sometimes, reliable transmission between a certain pair of CR nodes may require high power, resulting in heavy interference to the primary system. In this case, conventional direct transmis-

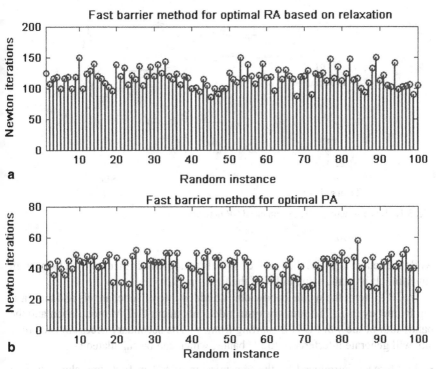

Fig. 3.10 Number of Newton iterations for convergence for 100 channel realizations. $N = 64, K = 4, K_1 = 2, L = 2$. **a** Fast barrier method for RA based on relaxation. **b** Fast barrier method for the optimal power allocation

sion is no longer permitted [11]. Instead, cooperative relay can not only weaken the effect of multipath fading, but also reduce the required transmission power by generating alternative paths between the source node and the destination [12, 13]. Thus, the interference cast to the primary system can be alleviated to some extent.

In this section, we study the basic framework of three-node relay systems and then propose our algorithms for resource allocation in more complex and practical multiuser relaying CR systems.

3.3.1   Three-Node Relay System

Consider a basic three-node relay-enhanced CR network, sharing the whole spectrum with a primary system, as shown in Fig. 3.11. The CR system adopts OFDM modulation. The relay operates in a time-division half duplex mode [14]. The transmission from the CR SN to its DN is on a time-frame basis and each frame consists of two time slots with equal duration. The source transmits signals while the relay and the receiver of the DN listen in the first time slot. Then the source remains silent

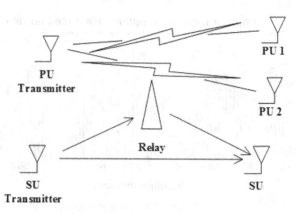

Fig. 3.11 Basic structure of relay enhanced CR networks

while the relay amplifies the received signals and forwards them to the DN over all subchannels with the same index as the first time slot.

In CR system, the whole bandwidth is divided into N subchannels and the starting frequency is f_0. The primary system does not adopt OFDM modulation and the bandwidth of the l th PU is B_l starting from f_l. The secondary access in each time slot will generate interference to PU bands, which can be calculated as

- Interference introduced by the transmission of SN in first time slot with unit power

$$I_{n,l}^{SP} = \int_{f_l - f_0 - (n-1/2)W/N}^{f_l + B_l - f_0 - (n-1/2)W/N} g_{n,l}^{SP} \phi(f) df, \qquad (3.24)$$

- Interference introduced by the transmission of RN in second time slot with unit power

$$I_{n,l}^{RP} = \int_{f_l - f_0 - (n-1/2)W/N}^{f_l + B_l - f_0 - (n-1/2)W/N} g_{n,l}^{RP} \phi(f) df, \qquad (3.25)$$

where $\phi(f)$ is the baseband PSD of OFDM signal.

Here, we investigate the capacity for the relay channel, both for AF and DF protocol, which are the most frequently adopted in practical systems. Generally, there are two common scenarios for two-hop relaying transmission, namely, with diversity and without diversity. The former means that the signal from the direct path (SN to DN) is taken into account, where the DN combines the signals from the direct path and the relayed signal by maximal ratio combining (MRC). The latter signifies that the DN only receives the relayed signal from the RN.

- Case A: AF protocol

We assume that the SN transmits with power $p_{s,n}$ on the nth subchannel in the first hop and the relay amplifies the received signals on the nth subchannel with power $P_{r,n}$ in the second hop. The amplification factor on the nth subchannel is

$$\beta_n = \sqrt{\frac{P_{r,n}}{P_{s,n}\left|h_{sr,n}\right|^2 + N_r B + I_n^{PR}}},$$

where $h_{sr,n}$ is the channel gain from the SN to RN, and N_r is the PSD of AWGN at the CR relay. The interference generated by transmission of PUs on the nth subchannel at the RN is denoted by I_n^{PR}, which can be measured as noise at the RN. After MRC, the SNR for the nth subchannel at the DN is given by [14]

$$\gamma_n = \frac{P_{s,n}\left|h_{rd,n}\right|^2 \beta_n^2 \left|h_{sr,n}\right|^2}{N_d B + I_n^{PD} + (N_r B + I_n^{PR})\beta_n^2 \left|h_{rd,n}\right|^2} + \frac{P_{s,n}\left|h_{sd,n}\right|^2}{N_d B + I_{PD}^n}$$

$$= \frac{P_{s,n} a_n P_{r,n} b_n}{1 + p_{s,n} a_n + p_{r,n} b_n} + P_{s,n} c_n.$$

If the signal from direct path is not considered, namely without diversity, the SNR at the receiver of DN of the nth subchannel can be otherwise calculated by [15]

$$\gamma_n = \frac{P_{s,n} a_n P_{r,n} b_n}{1 + p_{s,n} a_n + p_{r,n} b_n}.$$

Here, $h_{rd,n}$ and $h_{sd,n}$ are the channel gain from the RN to the DN and from the SN to the DN, respectively. The interference introduced by PUs at the DN on the nth subchannel, I_n^{PD}, is regarded as noise and can be measured by the DN. Besides, $a_n = \dfrac{\left|h_{sr,n}\right|^2}{(N_r B + I_n^{PR})}$, $b_n = \dfrac{\left|h_{rd,n}\right|^2}{(N_d B + I_n^{PD})}$ and $c_n = \dfrac{\left|h_{sd,n}\right|^2}{(N_d B + I_n^{PD})}$ are the normalized SNRs of the transmission link from SN to RN, RN to DN and SN to DN in the relaying CR system.

Consequently, the transmission rate on the nth subchannel is given by

$$r_n^{AF} = \frac{1}{2}\log\left(1 + \frac{\gamma_n}{\Gamma}\right).$$

The rate of scaled by the factor 1/2 because the transmission takes two time slots.

- Case B: DF protocol

Assume the transmission powers at the SN and the RN are $P_{s,n}$ and $P_{r,n}$ over the nth subchannel, respectively, the relay is half-duplex and adopts DF protocol; that is, the SN transmits in the first time slot, then the relay forward the decoded signals

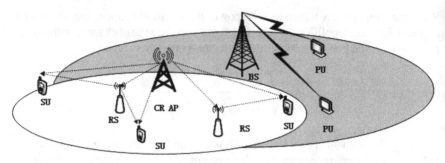

Fig. 3.12 System model of multiuser relaying systems

to the DN in the second time slot [16]. When the DN only receives signal from the RN, the channel capacity is limited by the minimum of the two hops [9]. Thus the transmission rate of the nth subchannel without diversity is

$$r_n^{DF} = \frac{1}{2} \min \left\{ \log \left(1 + \frac{p_{s,n} a_n}{\Gamma} \right), \log \left(1 + \frac{p_{r,n} b_n}{\Gamma} \right) \right\}.$$

When the direct path between the SN and the DN is taken into account, the DN combines the signals from the SN and the RN through MRC, which gives the rate of the nth subchannel as [16]

$$r_n^{DF} = \frac{1}{2} \min \left\{ \log \left(1 + \frac{p_{s,n} a_n}{\Gamma} \right), \log \left(1 + \frac{p_{r,n} b_n}{\Gamma} + \frac{p_{s,n} c_n}{\Gamma} \right) \right\}.$$

Different from the power allocation in conventional OFDM systems, power allocation in relaying systems is intended to determine the amount of power allocated to each subchannel at the source and the relay, where the total transmission power at the SN and the RN are both limited. The power allocation problem in the basic three-node relay systems has been investigated in [15, 17]. In this brief, the resource allocation, including subchannel assignment and power allocation, will be detailed for multiuser relaying CR system in the following section, where the power allocation in the basic two-hop system is involved. In other words, our method for resource allocation in multiuser relaying CR system can also be applied to the power allocation in three-node CR systems with minor modifications. Thus, we will not elaborate the power allocation in this case separately.

3.3.2 Multiuser Relaying CR Systems

Here, we consider a practical multiuser CR system enhanced by several relays, which coexists with a primary system. As illustrated in Fig. 3.12, the PUs are served by a base station and the SUs access via a CR AP. In the CR system, multiple relays with DF protocol are deployed to assist the communication between the AP and the SUs. We focus on the downlink of the CR system with K SUs enhanced by M relays. To reduce the deployment cost, relays are always established apart from each other to avoid the overlaps of coverage. That is to say, each SU can be covered by at most one relay according to its location. The relay that covers the kth SU is denoted by m_k. We assume that perfect channel state information is available at the transceivers of both the licensed system and the secondary system. The whole bandwidth is divided into N subchannels in the CR system, denoted as a set of $\mathcal{N} = \{1,\ldots,N\}$.

The relays are half-duplex, which means that they receive data in the first time slot and transmit them in the second time slot. To avoiding co-channel interference, only one node can transmit in a given time slot for each subchannel. Each relay re-encodes the received signal with the same codebook as that used by the source. The receiver of each SU uses MRC to combine the signals from the source and the relay, pertaining to the same message.

The CR system shares the whole bandwidth with L PUs, where PUs do not adopt OFDM modulation. For each time slot, the interference to the lth PU band is denoted by $I_{n,l}^{SP}$ and $I_{m_k,n,l}^{RP}$, implying the interference generated by the transmission of source (CR AP) on the nth subchannel and the interference caused by the transmission of the relay m_k on the nth subchannel with unit power, respectively. Mathematically, the interference power $I_{n,l}^{SP}$ and $I_{m_k,n,l}^{RP}$ can be calculated in the similar way as (3.24) and (3.25).

Denote $h_{k,n}^{SD}$, $h_{m_k,n}^{SR}$ and $h_{m_k,k,n}^{RD}$ as the channel gains of the nth subchannel from the CR AP to the kth SU, from the CR AP to the relay m_k and from the relay m_k to the kth SU, respectively. For simplicity, the normalized SNR dividing the SNR gap for each link is given by

$$a_{k,n} = \frac{\left|a_{m_k,n}^{SR}\right|^2}{\Gamma(\sigma_{m_k}^2 + I_{m_k}^{PR})}, b_{k,n} = \frac{\left|a_{m_k,k,n}^{RD}\right|^2}{\Gamma(\sigma_k^2 + I_k^{PS})}, c_{k,n} = \frac{\left|a_{k,n}^{SD}\right|^2}{\Gamma(\sigma_k^2 + I_k^{PS})}$$

Let $r_{1,k,n}$ be the transmission rate of the nth subchannel used by the kth SU assisted by the relay m_k,

$$r_{1,k,n} = \frac{1}{2}\min\left\{\log\left(1 + p_{s,k,n}a_{k,n}\right),\ \log\left(1 + p_{s,k,n}c_{k,n} + p_{r,k,n}b_{k,n}\right)\right\},$$

where $p_{s,k,n}$ and $p_{r,k,n}$ are the transmission power of the CR AP and the relay m_k on the nth subchannel used by the kth SU, respectively. On the other hand, if the relay does not forward, the achievable rate of the kth SU over the nth subchannel in this direct transmission mode is,

$$r_{2,k,n} = \frac{1}{2}\log\left(1 + p_{s,k,n}c_{k,n}\right).$$

It is easy to prove that the achievable rate of direct mode might be larger than that of relaying mode. Thus, in order to maximize the system capacity, appropriate transmission mode should be applied different SUs with different subchannels dynamically.

In order to maximize the SE of CR system, we try to maximize the sum rate of all SUs under the individual transmission power limit of the AP and the relays, while keeping the interference generated to each PU bands below its threshold. Here, a binary variable $t_{k,n}$ is introduced to indicate whether the relay m_k is active for the transmission of the kth SU over the nth subchannel. If $p_{r,k,n} \neq 0$, then $t_{k,n} = 1$; otherwise, $t_{k,n} = 0$. The subchannel assignment indices are denoted by the binary variables $p_{k,n}$'s and we have $p_{k,n} = 1$ only when the nth subchannel is allocated to the kth SU. Consequently, the RA task can be formulated as follows,

$$
\max_{\substack{t_{k,n}, \rho_{k,n}, \\ p_{s,k,n}, p_{r,k,n}}} \quad \sum_{k=1}^{K} \sum_{n=1}^{N} t_{k,n} \rho_{k,n} r_{1,k,n} + (1 - t_{k,n}) \rho_{k,n} r_{2,k,n}
$$

$$
s.t. \quad C1 \quad \sum_{k=1}^{K} \sum_{n=1}^{N} \rho_{k,n} p_{s,k,n} \leq P_s,
$$

$$
C2 \quad \sum_{k=1}^{K} \sum_{n=1}^{N} \rho_{k,n} p_{r,k,n} \leq P_r,
$$

$$
C3 \quad \sum_{k=1}^{K} \sum_{n=1}^{N} \rho_{k,n} I_{n,l}^{SP} p_{s,k,n} \leq I_l^{th}, l = 1, \ldots, L,
$$

$$
C4 \quad \sum_{k=1}^{K} \sum_{n=1}^{N} \rho_{k,n} I_{k,n,l}^{RP} p_{r,k,n} \leq I_l^{th}, l = 1, \ldots, L, \qquad (3.26)
$$

$$
C5 \quad \rho_{k,n} \in \{0,1\}, \forall n, \forall k,
$$

$$
C6 \quad \sum_{k=1}^{K} \rho_{k,n} = 1, \forall n,
$$

$$
C7 \quad p_{s,k,n} \geq 0, p_{r,k,n} \geq 0, \forall n, \forall k,
$$

$$
C8 \quad t_{k,n} = \{0,1\}, \forall n, \forall k,
$$

where P_s and P_r are the maximum transmission power of the CR AP and relays. C3 and C4 describe the interference limit for PUs in each time slot, where I_l^{th} is the interference power limit of the lth PU. C5 and C6 imply that each subchannel can only be occupied by one SU in the CR system.

Obviously, problem (3.26) defines a mixed integer programming problem, since it involves binary variables $t_{k,n}$'s and $\rho_{k,n}$'s, together with continuous variables $p_{s,k,n}$'s and $p_{r,k,n}$'s. Generally, it is hard to solve the problem because the integer constraints generate an exponential complexity. Thus, we develop a two-stage method to address it, which carries out the subchannel assignment and power distribution in turn. During such process, the transmission mode of each subchannel can be indirectly determined.

- Subchannel allocation

The achievable rate on the nth subchannel used by the kth SU can be unified into

$$
r_{k,n} = \begin{cases} r_{1,k,n} & p_{r,k,n} \neq 0 \\ r_{2,k,n} & p_{r,k,n} = 0 \end{cases}.
$$

Proposition 3-1: The nth subchannel used by the kth SU that satisfies $c_{k,n} \geq a_{k,n}$ can achieve higher rate without relaying.

Proof If $c_{k,n} \geq a_{k,n}$, we have

$$\min\left\{\log\left(1+p_{s,k,n}a_{k,n}\right),\log\left(1+p_{s,k,n}c_{k,n}+p_{r,k,n}b_{k,n}\right)\right\}$$
$$= \log\left(1+p_{s,k,n}a_{k,n}\right) \leq \log\left(1+p_{s,k,n}c_{k,n}\right).$$

On the other hand, direct transmission can eliminate the interference to the PUs in the second time slot. That is to say, in this case, direct transmission between the AP and the kth SU on the nth subchannel is preferred, since $r_{2,k,n} \geq r_{1,k,n}$ when $c_{k,n} \geq a_{k,n}$.

Note that $r_{2,k,n} = r_{1,k,n}$, when $p_{r,k,n} = 0$ and $c_{k,n} < a_{k,n}$. Accordingly, some subchannels with $c_{k,n} < a_{k,n}$ may be selected for direct transmission mode without relaying when the transmission power for the relays is not sufficient due to the power or interference limited. Thus, we can roughly classify the transmission mode into

$$r_{k,n} = \begin{cases} r_{1,k,n} & c_{k,n} < a_{k,n} \\ r_{2,k,n} & c_{k,n} \geq a_{k,n} \end{cases}, \tag{3.27}$$

where the accurate transmission mode of $c_{k,n} < a_{k,n}$ with $r_{k,n} = r_{1,k,n}$ can be further determined during power allocation procedure.

For subchannel assignment, the condition of each channel is evaluated by its maximum achievable rate. In CR systems, since a subchannel with higher SNR may generate more interference to PUs, the power limit and interference thresholds jointly set an upper bound of the maximum transmission power of each subchannel. Hence, similar to (3.15), we jointly consider the SNR of a subchannel and the interference to PUs thrown by it, developing a criterion of the maximum achievable rate of each subchannel,

$$r_{k,n}^M = \frac{1}{2}\begin{cases} \log\left(1+\min\left\{p_{s,k,n}^M a_{k,n}, p_{s,k,n}^M c_{k,n}+p_{r,k,n}^M b_{k,n}\right\}\right) & c_{k,n} < a_{k,n} \\ \log\left(1+p_{s,k,n}^M c_{k,n}\right) & c_{k,n} \geq a_{k,n} \end{cases}$$

where $r_{k,n}^M$ can be regarded as the maximum achievable rate over the nth subchannel used by the kth SU, $p_{s,k,n}^M$ and $p_{s,k,n}^M$ are the maximum possible power on the nth subcarrier used by the kth SU at the AP and the relay m_k, respectively,

$$p_{s,k,n}^M = \min\left\{P_s, \min_{l\in L}(I_{n,l}^{th}/I_{n,l}^{SP})\right\}, p_{r,k,n}^M = \min\left\{P_r, \min_{l\in L}(I_l^{th}/I_{k,n,l}^{RP})\right\}.$$

Preferably, each subchannel is allocated to the SU with the highest maximum achievable rate over it. This procedure terminates until all subchannels are consumed. Mathematically, we can determine the binary variables $\rho_{k,n}$'s according to the following equation,

$$\rho_{k,n} = \begin{cases} 1 & k = \underset{k}{\operatorname{argmax}}\ r_{k,n}^{M} \\ 0 & \text{otherwise} \end{cases}.$$

Recall that the variables $r_{k,n}$'s are roughly determined by (3.27), which will be further adjusted during power allocation.

- Power loading

After preliminary subchannel allocation and mode selection, power allocation among subchannels can best capitalize on the limited resource to achieve the potential capacity of system. According to (3.27), we resort and classify all subchannels into two categories as follows. Denote $\mathcal{N}_1 = \{1,\ldots,N_1\}$ as the subset of subchannels with $c_{k,n} < a_{k,n}$, and $\mathcal{N}_2 = \{N_1+1,\ldots,N\}$ as the subset of subchannels with $c_{k,n} \geq a_{k,n}$. Then the power allocation problem can be written into

$$\max_{p_{s,n},p_{r,n}} \sum_{n\in\mathcal{N}_1} r_{1,n} + \sum_{n\in\mathcal{N}_2} r_{2,n}$$

$$s.t.\ C1\quad \sum_{n=1}^{N} p_{s,n} \leq P_s,$$

$$C2\quad \sum_{n=1}^{N} p_{r,n} \leq P_r,$$

$$C3\quad \sum_{n=1}^{N} I_{n,l}^{SP} p_{s,n} \leq I_l^{th} = l = 1,\ldots,L,$$

$$C4\quad \sum_{n=1}^{N} I_{n,l}^{RP} p_{r,n} \leq I_l^{th} = l = 1,\ldots,L,$$

$$C5\quad p_{s,n} \geq 0, p_{r,n} \geq 0, \forall n,$$

(3.28)

where all the subscripts k's are omitted because each subcarrier is associated with a fixed SU.

The max-min formulation in the objective function in (3.28) can be addressed by introducing extra auxiliary variables z_n's, $\forall n$. The power allocation problem can be transformed into

$$\max_{p_{s,n},p_{r,n},z_n} \frac{1}{2}\sum_{n=1}^{N} \log(1+z_n)$$

$$s.t.\quad z_n \leq p_{s,n} h_n^{SR}, n \in \mathcal{N}_1$$

$$z_n \leq p_{s,n} h_n^{SD} + p_{r,n} h_n^{RD}, n \in \mathcal{N}_1$$

$$z_n \leq p_{s,n} h_n^{SD}, n \in \mathcal{N}_2$$

$$z_n \geq 0, \forall n$$

$$C1 \sim C6 \text{ in (3.28)}.$$

(3.29)

Indeed, the feasible set of (3.29) always covers that of (3.28) and the two problems share the same optimal solution. It can be proved that (3.29) is jointly convex with respect to the optimization variables $\{p_{s,n}, p_{r,n}, z_n\}$'s.

Although standard barrier method is an effective way to solve the above convex optimization problem, its high complexity is unacceptable in practical scenarios. Inspired by the idea of fast barrier method with lower complexity introduced in the previous section, we further explore the structure of (3.29) to figure out whether it is possible to speed up the computation of Newton step.

First, for the barrier method [1], the approximated minimization problem of (3.29) is given by

$$\min_{x} \psi_t(x) = -\frac{t}{2} \sum_{n=1}^{N} \log(1 + z_n) + \phi(x). \tag{3.30}$$

As the parameter t increases, the optimal solution to (3.30) will approximate to the optimal solution of the original problem more and more accurately. Denote the optimization variable as

$$x = \left\{ p_{s,1}, p_{r,1}, z_1, \ldots, z_{N_1}, p_{s,N_1+1}, z_{N_1+1}, \ldots, z_N \right\},$$

and the barrier function

$$\phi(x) = -\sum_{n \in \mathcal{N}_1} \log\left(p_{s,n} c_n + p_{r,n} b_n - z_n\right) - \sum_{n \in \mathcal{N}_1} \log\left(p_{s,n} a_n - z_n\right)$$

$$- \sum_{n \in \mathcal{N}_2} \log\left(p_{s,n} c_n - z_n\right) - \log f_s - \log f_r - \sum_{l=1}^{L} \log f_{s,l} - \sum_{l=1}^{L} \log f_{r,l}$$

$$- \sum_{n=1}^{N} \log p_{s,n} - \sum_{n \in \mathcal{N}_1} \log p_{r,n} - \sum_{n=1}^{N} \log z_n,$$

where

$$f_s = P_s - \sum_{n=1}^{N} p_{s,n}, \quad f_r = P_r - \sum_{n=1}^{N_1} p_{r,n},$$

$$f_{s,l} = I_l^{th} - \sum_{n=1}^{N} p_{s,n} I_{n,l}^{SP}, \quad f_{r,l} = I_l^{th} - \sum_{n=1}^{N_1} p_{r,n} I_{n,l}^{RP}, l = 1, \ldots, L.$$

To solve the unconstrained minimization problems (3.30), Newton method is always employed [1]. With a given parameter t, Newton step Δx is given by

$$\nabla^2 \psi_t(x) \Delta x = -\nabla \psi_t(x). \tag{3.31}$$

Supposing we compute Newton step via direct matrix inversion of (3.31), it will cost $O(N^3)$, which is evidently unaffordable for practical systems. The Hessian of $\psi_t(x)$ is as follows:

$$\nabla^2 \psi_t(x) = \begin{bmatrix} D_1 & & \\ & \ddots & \\ & & D_N \end{bmatrix} + \frac{\nabla f_s \nabla f_s^T}{f_s^2}$$

$$+ \sum_{l=1}^{L} \frac{\nabla f_{s,l} \nabla f_{s,l}^T}{f_{s,l}^2} + \frac{\nabla f_r \nabla f_s^T}{f_r^2} + \sum_{l=1}^{L} \frac{\nabla f_{r,l} \nabla f_{r,l}^T}{f_{r,l}^2}$$

where

$$D_n = \begin{bmatrix} 1/p_{s,n}^2 & & \\ & 1/p_{r,n}^2 & \\ & & 1/z_n^2 + t/(1+z_n)^2 \end{bmatrix}$$
$$+ \omega_n^2 \begin{bmatrix} c_n^2 & c_n b_n & -c_n \\ c_n b_n & b_n^2 & -b_n \\ -c_n & -b_n & 1 \end{bmatrix} + \upsilon_n^2 \begin{bmatrix} a_n^2 & 0 & -a_n \\ 0 & 0 & 0 \\ -a_n & 0 & 1 \end{bmatrix}, n \in \mathcal{N}_1,$$

and

$$D_n = \begin{bmatrix} 1/p_{s,n}^2 & \\ & 1/z_n^2 + t/(1+z_n)^2 \end{bmatrix} + \omega_n^2 \begin{bmatrix} c_n^2 & -c_n \\ -c_n & 1 \end{bmatrix}, n \in \mathcal{N}_2,$$

where

$$\omega_n = \begin{cases} \dfrac{1}{p_{s,n} h_n^{SD} + p_{r,n} h_n^{RD} - z_n}, n \in \mathcal{N}_1 \\[3ex] \dfrac{1}{p_{s,n} h_n^{SD} - z_n}, \quad n \in \mathcal{N}_2 \end{cases}$$

$$\upsilon_n = \begin{cases} \dfrac{1}{p_{s,n} h_n^{SR} - z_n}, n \in \mathcal{N}_1 \\[2ex] 0, \qquad\qquad n \in \mathcal{N}_2. \end{cases}$$

Accordingly, the Hessian of $\psi_t(x)$ can be decomposed into a block-arrow matrix D and $2L+2$ one-rank matrices. Thus, the idea of fast computation of Newton step introduced in previous sections is applicable in this problem. The procedure is similar to that in Table 3.3, which is elaborated in [18]. In this case, the complexity to work out the optimal solution can be roughly measured by $O(M^2 N)$.

3.3.3 Numerical Results

In simulation, we consider a multiuser OFDM-based CR relaying network, where all SUs and relays are located at two concentric ring-shaped discs around an AP. The outer boundary has a radius of 600 m and the inner boundary has a radius of 400 m. There are four relays in the CR system, uniformly located at the boundary of inner circle, and all SUs distributed in the outer ring uniformly, each of which is covered by a certain relay. PUs are located around a BS within a circle with radius of 1 km and the PSD of PU signals is an elliptically filtered white noise process. The channel suffers from frequency selective fading and the path loss exponent is 4. Assume the variance of shadowing effect is 10 dB and the amplitude of multipath fading follows Rayleigh distribution. The bandwidth of each PU is randomly generated by uniform distribution and the maximum value is $2W/3L$ with the total bandwidth W. The noise power is set to 10^{-13} W.

To evaluate the performance of our proposed RA algorithm, other three schemes are introduced for comparison:

1. All relaying w/o diversity (RND). The direct link is not taken into account that all signals are forwarded by relays.
2. Fixed direct mode (FDM). The conventional transmission mode from the CR AP to the SUs is employed without the assistance of relays.
3. Equal power allocation (EPC). For our proposed transmission mode, the power is equally distributed among subchannels.

Figure 3.13 illustrates the sum rate of all SUs as a function of the transmission power limit at (a) the CR AP and (b) the relays, respectively. There are 32 OFDM subchannels, eight SUs and two PUs. The interference threshold of each PU band is 10^{-10} W. We find that the growth of system capacity slows down with the increase of power budget because more subchannels become interference limited as the transmission power gets larger. Moreover, the RND scheme can achieve about 90 % of our proposal, where the loss of capacity is attributed to its fixed relaying transmission mode. Since the relay does not participate in transmission, the FDM fails to exploit the spatial diversity, resulting in its poor performance. Furthermore, we investigate the curves of the sum rate for different numbers of subchannels versus the power limit P_s and P_r in Figs. 3.14 and 3.15. For a given power budget, the capacity increases when the number of subchannels gets larger, which implies that channel diversity contributes to significant gain of system capacity. Also, it is shown that the fixed power allocation scheme EPC can only achieve less than 50 % of our proposal, which confirms the necessity of dynamic power allocation.

Figure 3.16 describes the sum rate of CR system versus the interference threshold for different numbers of PUs. There are eight SUs and 32 subchannels with fixed power limit at the AP and relays of $P_s = P_r = 1 W$. Three different cases in terms of the number of PUs are investigated: $L = 2$, $L = 4$ and $L = 6$. As illustrated in Fig. 3.16, the sum rate of the SUs increases with the growth of interference threshold until the subchannels become power limited for each case. On the other hand, with a given interference power limit, more, more PUs result in a relatively

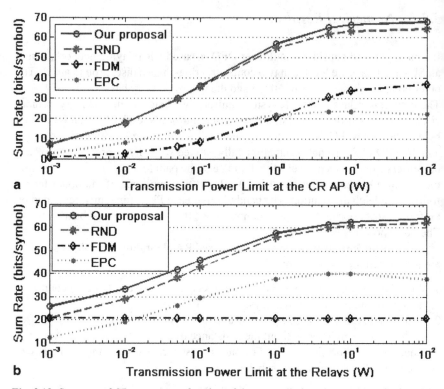

Fig. 3.13 Sum rate of CR system as a function of the power limit at CR AP (**a**) and relays (**b**). $N = 32$, $K = 8$, $L = 2$, $I_l^{th} = 10^{-13}\,W$ and (**a**) $P_r = 1\,W$, and (**b**) $P_s = 1\,W$

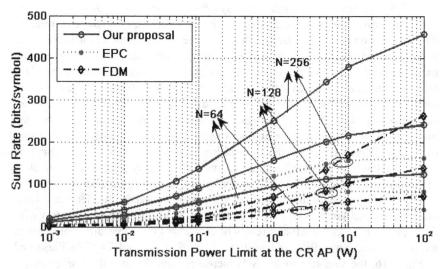

Fig. 3.14 Sum rate of CR system versus power limit at CR AP for different number of subchannels. $K = 8, L = 2$, $I_l^{th} = 10^{-13}\,W$ and $P_r = 1\,W$

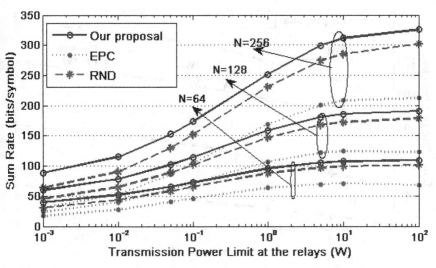

Fig. 3.15 Sum rate of CR system versus power limit at relays for different number of subchannels. $K = 8$, $L = 2$, $I_I^{th} = 10^{-13}\,W$ and $P_s = 1\,W$

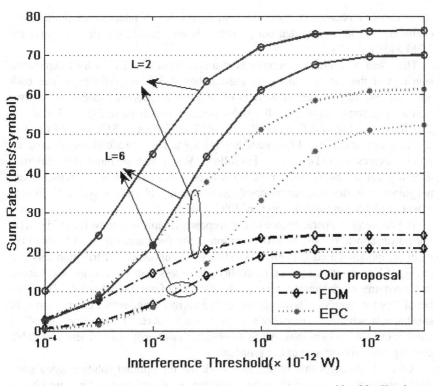

Fig. 3.16 Sum rate of CR system as a function of interference threshold. $N = 32$, $K = 8$ and $P_s = P_r = 1\,W$

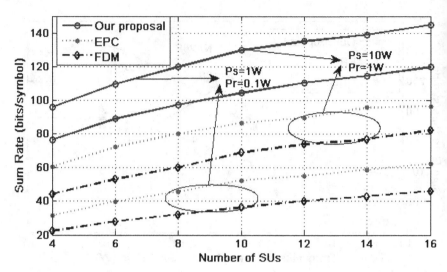

Fig. 3.17 Sum rate of CR system as a function of the number of SUs. $N = 64$, $L = 2$ and $I_l^{th} = 10^{-13}$ W

lower capacity because of more frequent occurrence of interference limited sub-channels. It is apparently that our proposal shows significant superiority over the FDM and the EPC schemes.

The effect of multiuser diversity is also studied in Fig. 3.17, which depicts the sum rate of the CR system versus the number of SUs for different power budgets. The number of subchannels is 64 and the interference thresholds of two PUs are uniformly set to 10^{-13} W. Two cases are considered in Fig. 3.17, that are, $P_s = 10\,W, P_r = 1\,W$ and $P_s = 1\,W, P_r = 0.1\,W$. The number of SUs varies from 4 to 16. As observed in Fig. 3.17, more SUs gives rise to the increase of overall capacity for all schemes. It can be explained as follows: When there are more SUs, multiuser diversity makes a subchannel more likely to be allocated to an SU with high channel gain over it. Besides, our proposal can thoroughly take advantage of the limited power, achieving more than 150 % of EPC scheme.

The convergence performance of our proposed scheme is manifested in Fig. 3.18. For power allocation, the Newton iteration in the fast barrier method leads to the main computational cost. Figure 3.18 shows the cumulative distribution function (CDF) of Newton iterations with a duality gap of less than 10^{-3}. It can be observed that it requires more Newton iterations for the case of more subchannels. The number of Newton iterations, however, varies in a narrow range with a given number of subchannels, while the increase of subchannels will not result drastic increase of the number of Newton iterations. All these observations verify the efficient and stable convergence performance of our proposal.

Although the algorithm in also can work out the optimal solution using standard interior point method (Standard), the computational cost is high, which can

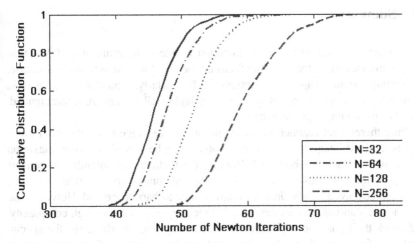

Fig. 3.18 CDF of number of Newton iterations for convergence

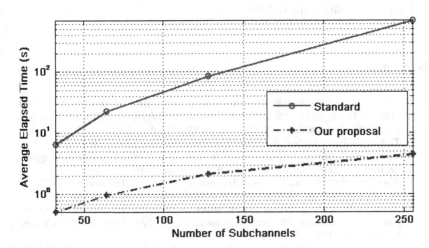

Fig. 3.19 Average time cost as a function of the number of subchannels

be intuitively reflected by its elapsed time for convergence. Figure 3.19 gives the average time cost as a function of subchannels over 1000 realizations. The elapsed time is counted by in-built *tic-toc* function in *Matlab*. Obviously, the time cost of our proposal is much less than that of the standard technique employed in [19], which further verifies the remarkable convergence performance of our proposed algorithm.

From the results in Figs. 3.13–3.19, we can conservatively conclude that our proposed RA scheme is effective and efficient for possible applications in practical wireless systems.

3.4 Summary

In this chapter, specified resource allocation problems in terms of OFDM-based CR systems are elaborated, with the focus on the SE of the CR system. In this case, maximizing the sum capacity is preferable. Particularly, typical RA problems in single SU system, multiuser CR system and relaying CR system are recommended, as well as the methods to solve them.

While there is subchannel allocation involved, we suggest that the RA formulated as a mixed integer programming problem can be solved by either relaxation method or two-stage method [20]. Both of the methods are intended to convert or simplify the intractable problems into standard convex optimization problems, which can be solved by standard techniques, such as barrier method. However, the drawback of standard convex optimization techniques, known as high complexity, holds back their possible application in practical systems. Fortunately, the special structures of these problems allow the significant improvement of standard method, based on which we propose fast algorithms to work out (near) optimal solutions. Our proposed algorithms break through the bottleneck of the standard method and show great potential for applications.

References

1. S. P. Boyd and L. Vandenberghe, *Convex Optimization*. Cambridge University Press, 2004.
2. S. Wang, F. Huang, and Z. Zhou, "Fast power allocation algorithm for cognitive radio networks," *IEEE Commun. Lett.*, vol. 15, no. 8, pp. 845–847, Aug. 2011.
3. C. D. Meyer, Matrix Analysis & Applied Linear Algebra. SIAM Press, 2000.
4. Y. Zhang and C. Leung, "Resource allocation in an OFDM-based cognitive radio system," *IEEE Trans. Commun.*, vol. 57, no. 7, pp. 1928–1931, July 2009.
5. S. Wang, "Efficient resource allocation algorithm for cognitive OFDM systems," *IEEE Commun. Lett.*, vol. 14, no. 8, pp. 725–727, Aug. 2010.
6. M. Ge and S. Wang, "Fast optimal resource allocation is possible for multiuser OFDM-based cognitive radio networks with heterogeneous services," *IEEE Trans. Wireless Commun.*, vol. 11, no. 4, pp. 1500–1509, Apr. 2012.
7. W. Yu and J. Cioffi, "FDMA capacity of Gaussian multiple-access channels with ISI," *IEEE Trans. Commun.*, vol. 50, no. 1, pp. 102–111, Jan. 2002.
8. S. Wang, Z.-H. Zhou, M. Ge and C. Wang, "Resource allocation for heterogeneous cognitive radio networks with imperfect spectrum sensing," *IEEE J. Sel. Areas Commun.*, vol. 31, no. 3, pp. 464–475, 2013.
9. T. Cover and A. Gamal, "Capacity theorems for the relay channel," *IEEE Trans. Inf. Theory*, vol. 25, no. 5, pp. 572–584, Sep. 1979.
10. A. Nosratinia, T. Hunter, and A. Hedayat, "Cooperative communication in wireless networks," *IEEE Commun. Mag.*, vol. 42, no. 10, pp. 74–80, Oct. 2004.
11. D. Zhang, Y. Wang, and J. Lu, "QoS aware relay selection and subcarrier allocation in cooperative OFDMA systems," *IEEE Commun. Lett.*, vol. 14, no. 4, pp. 294–296, Apr. 2010.
12. A. Sendonaris, E. Erkip and B. Aazhang, "User cooperation diversity part I: System description," *IEEE Trans. Commun.*, vol. 51, no. 11, pp. 1927–1938, Nov. 2003.

13. A. Sendonaris, E. Erkip and B. Aazhang, "User cooperation diversity part II: Implementation aspects and performance analysis," *IEEE Trans. Commun.*, vol. 51, no. 11, pp. 1939–1948, Nov. 2003.
14. J. Laneman, D. Tse and G. Wornell, "Cooperative diversity in wireless networks: Efficient protocols and outage behavior," *IEEE Trans. Inform. Theory*, vol. 50, no. 12, pp. 3062–3080, Dec. 2004.
15. S. Wang, F. Huang, M. Ge and C. Wang, "Optimal power allocation for OFDM-based cooperative relay cognitive radio networks," in *Proc. IEEE ICC'12*, pp. 1651–1655, June 2012.
16. Q. Zhang, J. Zhang, C. Shao, Y. Wang, P. Zhang and R. Hu, "Power allocation for regenerative relay channel with Rayleigh fading," in *Proc. IEEE VTC'04*, pp. 1167–1171, May 2004.
17. Z. Shu, W. Chen, "Optimal power allocation in cognitive relay networks under different power constraints," in *Proc. IEEE WCNIS*, pp. 647–652, June 2010.
18. S. Wang, M. Ge, C. Wang, "Efficient Resource Allocation for Cognitive Radio Networks with Cooperative Relays," *IEEE J. Sel. Areas Commun.*, vol. 31, no. 11, pp. 2432–2441, Nov. 2013.
19. D. Bharadia, G. Bansal, P. Kaligineedi, and V. Bhargava, "Relay and power allocation schemes for OFDM-based cognitive radio systems," *IEEE Trans. Wireless Commun.*, vol. 10, no. 9, pp. 2812–2817, Sep. 2011.
20. S. Wang, Z.-H. Zhou, M. Ge and C. Wang, "Resource Allocation for Heterogeneous Multiuser OFDM-based Cognitive Radio Networks with Imperfect Spectrum Sensing," in *Proc. IEEE INFOCOM'12*, pp. 2264–2272, Mar. 2012.

Chapter 4
Energy-Efficient Resource Allocation in CR Systems

With the explosive growing demands for high data-rate wireless services, energy consumption is also increasing at an alarming rate nowadays. Consequently, it leads to a large amount of greenhouse gas and high operation expenditure for wireless service providers [1]. Recently, green radio is becoming increasingly important and navigates new directions for research activities, with emphasis on the energy-efficiency (EE) in wireless systems. An overview of the EE concerned in wireless communications is surveyed in [2], which recommends the technical roadmaps of several major international projects for energy-efficient wireless networks and investigates the state-of-the-art research on this topic. Particularly, energy-efficient resource allocation (RA) has been put on the agenda in both industry and academia, especially for the OFDM-based systems [3]. Different from the two conventional classes of dynamic RA in OFDM systems—rate adaptive and margin adaptive [4], energy-efficient RA is a special case where the objective is generally to maximize or minimize a certain metric of EE for a wireless system. The most popular one is called "*bits-per-Joule*", defined as the system throughput with unit power consumption

However, it is noteworthy that there is limited work on the energy-efficient RA of CR networks, which actually is extremely necessary for the CR scenarios. Besides the greenhouse crisis and operation expenditure involved, it also a prerequisite to achieve high utilization of the limited transmission power, which is needed to support the additional signal processing requirements compared to non-CR systems, such as spectrum sensing.

In this chapter, we mainly explore the features of energy-efficient RA, developing valid methods for several typical problems in CR systems.

4.1 EE Characteristics

As aforementioned, the system EE is evaluated by the metric "*bits-per-Joule.*" We consider an OFDM system with N subchannels. The total bandwidth is W. The transmission rate over each subchannel is given by the Shannon formula

© The Author(s) 2014
S. Wang, *Cognitive Radio Networks*, SpringerBriefs in Computer Science,
DOI 10.1007/978-3-319-08936-2_4

$$r_n = \frac{W}{N} \log_2(1 + p_n h_n),$$

where p_n and h_n are the transmission power and the normalized SNR of the nth subchannel. Ideally, the EE of this OFDM system in *bits/Joule* is given by

$$\eta_{EE} = \frac{W \sum_{n=1}^{N} \log_2(1 + p_n h_n)}{N \sum_{n=1}^{N} p_n}.$$

In practical systems, however, the circuit power also accounts for a part of total consumed energy, which should be considered in system design. Indeed, in the transmission mode, the circuit power incurred by signal processing and active circuit blocks, such as AD/DA converters, synthesizer and mixer. As discussed in [5, 6], the circuit power consumption can be modeled as one static part and one dynamic part based on the parameters of active links. For instance, the circuit consumed energy can be described by a linear function of system capacity,

$$P_c = P_s + \zeta \sum_{n=1}^{N} r_n,$$

where P_s is the fixed part of the circuit power and ζ is a constant denoting the dynamic power consumption of unit data rate. In this case, the EE of system can be formulated as

$$\eta_{EE} = \frac{W \sum_{n=1}^{N} \log_2(1 + p_n h_n) / N}{\sum_{n=1}^{N} p_n + P_c}.$$

Generally, higher η_{EE} signifies the higher efficiency of consumed energy. The characteristic of the EE metric can be illustrated by the curve of its maximum value versus transmission power.

In simulation, we assume the total bandwidth is 0.96 MHz with 64 OFDM subchannels and the noise power is $10^{-13}\ W$. For a given transmission power, the maximum EE is achieved by maximizing the throughput by water-filling method [7]. The path loss exponent is 4, the variance of shadowing effect is 10 dB and the amplitude of multipath fading is Rayleigh.

To analyze the difference between the EE in ideal case and practical systems, we depict the EE curves versus the total transmission power $\sum_{n=1}^{N} p_n$ for both case in Fig. 4.1. It is shown that the EE is monotone decreasing with the total transmission power. That is to say, without concerning the circuit power, lower transmission power will lead to higher EE, which implies that maximizing EE is equivalent to minimizing transmission power $\sum_{n=1}^{N} p_n$, known as margin adaptive [4]. On the other hand, with the participation of circuit power $P_c = 2.5 + 0.5 \sum_{n=1}^{N} r_n W$, the EE

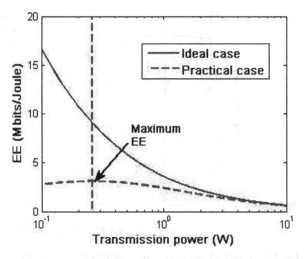

Fig. 4.1 EE versus transmission power for both ideal and practical cases

curve versus transmission power is not as simple as the ideal case. The circuit power breaks the monotonic relation between EE and $\sum_{n=1}^{N} P_n$, and the curve turns to a bell shape as shown in Fig. 4.1, which indicates the necessity and the importance of energy-efficient RA.

Additionally, the variation of circuit power, including static and dynamic part, may lead to different system decisions in EE optimization. To further understand the properties of the EE function, we also illustrate the EE curve versus transmission power with different parameters in circuit power in Fig. 4.2–4.3. When the dynamic circuit power factor ζ is fixed to 0.5 W/Mbit, the bell shaped EE curve goes down with the increase of static circuit power P_s in Fig. 4.2. Besides, the optimal power, denoted by P^*, to achieve the maximum EE on the curve, increases when P_s gets larger. As a counterpart, larger factor ζ also lowers the curve of EE, while the optimal power P^* remains unchanged with respect to different ζ's, as shown in Fig. 4.3. Such phenomena can be explained mathematically as follows,

$$\eta_{EE} = \frac{1}{\left(\sum_{n=1}^{N} p_n + P_s\right)\Big/ \sum_{n=1}^{N} r_n + \zeta},$$

$$\max_{p_n} \eta_{EE} \Leftrightarrow \min_{p_n} \frac{\sum_{n=1}^{N} p_n + P_s}{\sum_{n=1}^{N} r_n}.$$

It indicates that the optimal allocated power depends on the static power rather than the dynamic part, while higher circuit power will reduce the EE with a given transmission power. Taking advantage of this property, the energy-efficient RA can only consider the static part of circuit power, which will not result in different decisions.

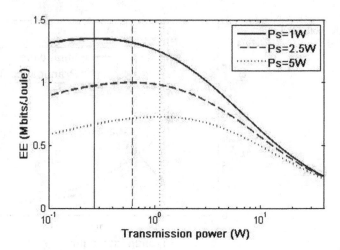

Fig. 4.2 EE versus transmission power for different static circuit powers

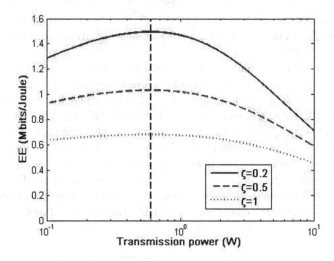

Fig. 4.3 EE versus transmission power for different parameters of dynamic circuit power

In most wireless communication systems, the total transmission power at the transmitter is usually limited. According to Fig. 4.1–4.3, the optimal EE also rely on the transmission power budget P_t. Particularly, if $P_t < P^*$, the optimal EE is not in the feasible regain, and EE is maximized by exhausting the maximum power P_t.

4.2   Energy-Efficient Power Allocation

With the knowledge of the main characters of EE function, we further investigate the energy-efficient power allocation problem under the consideration of many practical limitations, such as transmission power budget, interference threshold of PU and the traffic demand of the SU. Several possible methods are also proposed for solving this general problem.

4.2.1   General Problem

To investigate the nature of energy-efficient problem, the CR and primary systems are simplified to consist of only one SU and one PU. The case with multiple PUs can be easily extended with few modifications. The CR system is assumed to adopt OFDM modulation, sharing the whole bandwidth W with the primary system. The bandwidth of the nth subchannel spans from $f_0 + (n-1)\dfrac{W}{N}$ to $f_0 + n\dfrac{W}{N}$, where f_0 is the starting frequency and the spectrum of the PU spans from f_I to $f_I + W_I$.

The interference introduced to the PU by the transmission on the nth subchannel with unit power is I_n^{SP}, given by

$$I_n^{SP} = \int_{f_I - f_0 - (n-1/2)W/N}^{f_I + W_I - f_0 - (n-1/2)W/N} g_n^{SP} \phi(f) df,$$

where g_n^{SP} is the power gain from the CR AP to the PU's receiver. The interference to the nth subchannel cast by the transmission of the PU is denoted by I_n^{PS}, which can be regarded as noise and measured by the receiver of the SU. Thus, the SNR of the nth subchannel with unit power is

$$H_n = \frac{|h_n|^2}{N_0 \dfrac{W}{N} + I_n^{PS}}.$$

With the transmission power p_n on the nth subchannel, the achievable rate of the nth subchannel is given by

$$r_n = \frac{W}{N} \log_2(1 + p_n H_n).$$

For EE optimization, the power allocation result is irrelevant to the dynamic part of circuit power, which is a linear function of the throughput, as proved in the previous section. Hence, only the static circuit power P_s is considered in the optimization problem. Concerning the practical constraints in CR system, we assume the interference power to PU band is limited by its interference temperature limit. Besides, the overall throughput is desired to be no less than its minimal requirement. In short, the general form for energy-efficient power allocation can be expressed as

$$\max_{P_{s,n}, P_{r,n}} \quad \eta_{EE} = \frac{\sum_{n=1}^{N} r_n}{\sum_{n=1}^{N} p_n + P_s}$$

$$s.t. \quad C1 \quad \sum_{n=1}^{N} r_n \geq C_{\min}$$

$$C2 \quad \sum_{n=1}^{N} p_n \leq P_t$$

$$C3 \quad \sum_{n=1}^{N} p_n I_n^{SP} \leq I_{th}$$

$$C4 \quad P_n \geq 0, \forall_n, \tag{4.1}$$

where $C_{\min}$ is the minimal required throughput of CR system, P_t is the total transmission limit at the transmitter of CR system and Ith is the prescribed interference threshold of the PU band. Obviously, (4.1) defines a non-linear programming problem.

4.2.2   Bisection-Based Algorithm

Theorem 4.1 The function $\eta_{EE}(p)\cdot$ in (4.1) is strictly quasiconcave in variable p ($p = \{p_1, \ldots, p_N\}$). Besides, the $\eta_{EE}(p)$ is a quasiconvex optimization problem.

Proof Denote the superlevel sets of $\eta_{EE}(p)$ as

$$S_\alpha = \{p \in dom\,\eta_{EE}(p) \,/\, \eta_{EE}(p) \geq \alpha\}$$

for $\alpha \in \mathfrak{R}$. $\eta_{EE}(p)$ is strictly quasiconcave in p if and only if S_α is convex for any real number α [10]. S_α is equivalent to

$$S_\alpha = \left\{ p \in dom\,\eta_{EE}(p) \,\middle/\, \alpha \sum_{n=1}^{N} p_n - \sum_{n=1}^{N} W \log_2(1 + p_n H_n) + \alpha P_s \leq 0 \right\}\middle/N,$$

Which is obviously strictly convex in p. According to [10], we can conclude that unction $\eta_{EE}(p)$ in (4.1) is strictly quasiconcave. Since the inequality constraint functions in (4.1) are all convex, it can be further proved that (4.1) defines a quasiconvex optimization problem.

Denote $f(p) = -\eta_{EE}(p)$, which is quasiconvex in p. A general method to solve quasiconvex optimization problems relies on the representation of the sublevel sets of the quasiconvex objective function via a family of convex functions that satisfies

$$f(p) \leq t \Leftrightarrow \varphi_t(p) \leq 0,$$

Table 4.1 Bisection method for quasiconvex optimization

Algorithm 1
Given a tolerance $\epsilon > 0$ and an interval $[l, u]$, where $l \leq f^*, u \geq f^*$.
Repeat
$\tau = (l + u)/2$;
Solve the convex feasibility problem (4.2);
if problem (4.2) is feasible
$u = \tau$
else $l = \tau$;
Until $u - l \leq \epsilon$

i.e., the t-sublevel set of the quasiconvex function $f(p)$ is the 0-sublevel set of the convex function $\phi_t(p)$. Evidently, the convex function must guarantee the property that $\phi_s(p) \leq 0$ if $\phi_t(p) \leq 0, t \leq s$, which can be satisfied if $\phi_t(p)$ is a nonincreasing function of t for each p. To see that such representation always exist and not unique, we can take

$$\phi_t(p) = \begin{cases} 0 & f(p) \leq t \\ \infty, & \text{otherwise} \end{cases},$$

or

$$\phi_t(p) = -\sum_{n=1}^{N} r_n - t(\sum_{n=1}^{N} p_n + P_s).$$

Let η_{EE}^* denote the optimal solution to the problem (4.1), if the following problem

$$\begin{aligned} &\text{find} \quad p \\ &\text{s.t } \phi_\tau(p) \leq 0 \\ &\quad C1 \sim C4 \quad \text{in} \quad (4.1) \end{aligned} \qquad (4.2)$$

has a feasible solution, it implies $f^* \leq \tau$; otherwise, $f^* > \tau$.

Based on the observation discussed above, a simple bisection-based algorithm can be derived to solve the quasiconvex optimization problem. The outline of the outer iteration of bisection method is summarized in Table 4.1.

The interval is divided in two parts in each iteration, and the length of the interval after k iterations is $2^{-k}(u - 1)$, where $u - 1$ is the length of the initial interval. Hence, it requires exactly $\lceil \log_2((u - l)/\epsilon_b) \rceil$ iterations to converge to ϵ_b-optimality. In each iteration of the Bisection method, we need to solve a convex feasibility problem (4.2) with a given parameter τ. As discussed in [10], it is equivalent to solve a minimization problem by introducing a crucial indicator parameter z. According to [10], we can formulate the optimization problem to check the feasibility of the problem (4.2) as follows,

$$\min_{p,z} \quad z$$

$$s.t. \quad \phi_\tau(p) \le z$$

$$\sum_{n=1}^{N} W \log_2(1 + p_n H_n) / N \ge C_{\min} - z$$

$$\sum_{n=1}^{N} p_n \le P_t + z$$

$$\sum_{n=1}^{N} p_n I_n^{SP} \le I_{th} + z,$$

$$p_n \ge 0, \forall n. \tag{4.3}$$

The variable z can be interpreted as an upper bound of the maximum infeasibility of the inequalities as seen from (4.3). If the optimal solution of (4.3) is less than or equal zero, that is $z^* \le 0$, it means that at least one feasible solution to (4.2) exists. Furthermore, for a given τ that satisfies $f^* = \tau$, the solution $p_{k,n}$'s to (4.3) is optimal for the (4.1) if and only if $z^* = 0$ for (4.3). Thus, we can obtain the optimal resource allocation $p_{k,n}$'s by solving (4.3) in the last iteration with the solution of $z^* \le 0$.

As discussed above, we need to solve (4.3) for a given parameter τ in each iteration. If the representation function $\phi_t(p)$ is convex in p, which generally exists, (4.3) defines a convex optimization problem in $\{p, z\}$. Since the objective function is linear, the inequality constraints functions are all convex, and the equality functions are affine, it can be solved by standard convex optimization techniques.

4.2.3   Fractional Programming

Another characteristic of the EE function can be described as follows: η_{EE} is a nonlinear fractional function with concave numerator and linear (convex) denominator. Such special structure allows the utilization of fractional programming method [15].

First, a new optimization problem is defined as

$$\max_{p,\alpha} \quad g(p,\alpha) = \sum_{n=1}^{N} r_n - \alpha \left(\sum_{n=1}^{N} p_n + P_s \right)$$

$$s.t. \quad \sum_{n=1}^{N} r_n \ge C_{\min}$$

$$\sum_{n=1}^{N} p_n \le P_t$$

$$\sum_{n=1}^{N} p_n I_n^{SP} \le I_{th},$$

$$p_n \ge \forall n, \tag{4.4}$$

Table 4.2 Fractional programming

Algorithm 2
Initialization: parameter $\alpha = \alpha_0 > 0$, $F(\alpha) = \infty$ and tolerance $\delta > 0$.
While $|F(\alpha)| > \delta$
Solve problem (4.4) and obtain the optimal solution p^*
Calculate $F(\alpha)$
Update $\alpha = \eta_{EE}(p^*)$
Endwhile
Return p^* and α

where α is a positive parameter. Problem (4.1) and (4.4) share the same feasible region S. With a certain parameter α, the optimal value and solution of the problem (4.4) are $F(\alpha) = \min_p \{g(p, \alpha) \mid p \in S\}$, $f(\alpha) = \arg\min_p \{g(p, \alpha) \mid p \in S\}$, respectively. The relationship between problem (4.1) and (4.4) is given by the following lemma, which has been proved in [16].

Lemma 4-1: The optimal solution $p*$ to problem (4.4) at $\alpha*$ achieves the optimal value of problem (4.1) if and only if

$$F = (\alpha*) = 0,$$

which indicates that at the optimal parameter $\alpha*$, the optimal solution to (4.4) is also the optimal solution to (4.1). Therefore, the original problem can be solved by finding the optimal power allocation of the (4.4) for a given α and then updating α until the condition in **Lemma 4-1** is satisfied. The procedure of algorithm is detailed in Table 4.2.

In each iteration of updating parameter α, we have to solve the problem (4.4), which is a convex optimization problem and can be solved by standard techniques.

4.2.4   Hypograph-Based Algorithm

Since the original problem is difficult to solve directly, we turn to its equivalent transformation via its hypograph form [10], which is given by

$$\max_{p,y} \quad y$$

$$s.t. \quad n_{EE}(p) \geq y$$

$$\sum_{n=1}^{N} W \log_2(1 + p_n H_n) / N \geq C_{\min}$$

$$\sum_{n=1}^{N} p_n \leq P_t$$

$$\sum_{n=1}^{N} p_n I_n^{S,P} \leq I_{th},$$

$$y \geq 0, p_n \geq 0, \forall n.$$

Here, we try to maximize y over the hypograph of $\eta_{EE}(p)$ under the constraints in the original problem. The optimal power allocation $p*$ to the above problem is identical with the optimal solution to the original problem.

Furthermore, the first constraints in the equation above can be further converted into a convex form:

$$y\left(\sum_{n=1}^{N} p_n + P_s\right) - \sum_{n=1}^{N} r_n \leq 0,$$

which makes the hypograph problem convex. Rewrite the problem into

$$\max_{p,y} \quad y$$

$$s.t. \quad y\left(\sum_{n=1}^{N} p_n + P_s\right) - \sum_{n=1}^{N} r_n \leq 0$$

$$\sum_{n=1}^{N} W \log_2(1 + p_n H_n)/N \geq C_{min}$$

$$\sum_{n=1}^{N} p_n \leq P_t$$

$$\sum_{n=1}^{N} p_n I_n^{SP} \leq I_{th},$$

$$y \geq 0, p_n \geq 0, \forall n$$

It can be dealt with convex optimization techniques.

4.2.5 Performance Evaluation

The three methods discussed above are all effective to solve the considered energy-efficient power allocation problem, each of which has its advantages and drawbacks. Here, the performance of the algorithms is evaluated from two aspects-achievable EE and convergence performance via several experiments. In simulation, we assume the whole bandwidth 0.96 MHz is divided into 64 OFDM subchannels in the CR system. The receiver is randomly located in the circle within 0.5 km from its transmitter. The path loss exponent is 4, the variance of shadowing effect is 10dB and the amplitude of the multipath fading is Rayleigh. The PU band is randomly generated by uniform distribution with the maximum value of $2W/3L$. The noise power is 10^{-13} W and the interference threshold is 5×10^{-12} W.

Figure 4.4 shows the average EE of CR system as a function of the transmission power limit at the transmitter. The minimal rate requirement is set to 0.6 Mbps. The three algorithms almost achieve the same EE and the sight difference can be attributed to the algorithm accuracy. This result is coincident with the theoretical analysis above that the three algorithms can all work out the optimal solution of the considered problem. Another phenomenon in Fig. 4.4 is that the EE increases with the growth of power limit until the power budget is large enough, which further

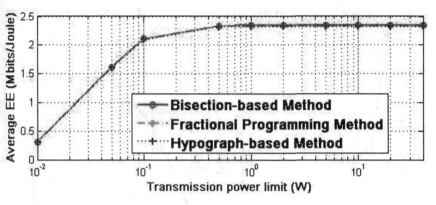

Fig. 4.4 The EE of CR system as a function of the transmission power limit

verifies the conclusion given in in Sect. 4.1. Recall the relationship between EE and the transmission power given in Fig. 4.3, when the power limit is relatively small, the maximum EE is always achieved by exhausting the maximum power. The maximum EE can be achieved at the optimal point and remains unchanged if the transmission power is large enough.

The convergence performance is illustrated in Fig. 4.5 and Table 4.3. For the first two methods, the outer iteration for convergence is shown in Fig. 4.5 and the convex optimization problem in the inner loop is solved by the barrier method. The midpoint of bisection method during each iteration is denoted by τ, which roughly requires nine iterations to converge, while the fractional programming converges within five outer iterations. Besides, the elapsed time for convergence of three methods is listed in Table 4.3, where the hypograph form of the considered problem in the last method is also addressed by the barrier method. We find that the hypograph-based method is more efficient than the other two ones.

From the above observation, it seems that the bisection-based method is the least preferred, since it generates more computational complexity. However, we also need to investigate the inherent advantages and limits of these methods. The bisection-based method is simple but effective for a series of quasiconvex optimization problems, while the other two methods impose more strict requirements on the problem. The fractional programming methods can only be applied to the case that the objective function of maximization has concave numerator and convex denominator with convex feasible regain. If the problem has a convex hypograph form, the last method is a better fit. After all, the considered problem is just a general form of energy-efficient power allocation, and there are more constraints and adjustments in practical systems, which may result in more intricate optimization problem. In a nutshell, different conditions need different methods.

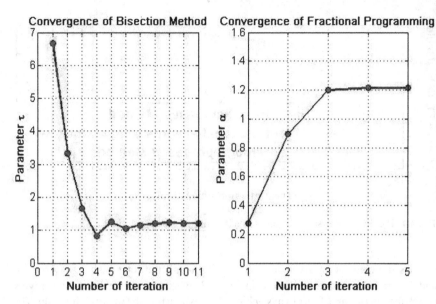

Fig. 4.5 Convergence performance of Bisection based method and fractional programming

Table 4.3 Time consumption for convergence (*seconds*)

N (Number of subchannels)	32	64	96	128
Bisection-based method	0.054	0.134	0.290	0.389
Fractional programming	0.021	0.027	0.034	0.047
Hypograph-based methods	0.007	0.009	0.010	0.019

4.3   Multiuser CR Systems

Based on the idea of energy-efficient power allocation discussed in the last section, we apply it into a practical multiuser CR system. Consider the downlink of an OFDM-based CR system with K SUs served by an AP, coexisting with L PUs in a primary system in the spectrum sharing manner. The whole bandwidth W is divided into N subchannels in the CR system, denoted by $N = \{1,\dots,N\}$. We assume that perfect knowledge of CSI is available at the transceivers in both CR and primary systems. Note that the results obtained by assuming instantaneous EE with perfect CSI can serve as an upper bound on the achievable EE with channel estimation errors.

4.3.1   Problem Formulation

Since the CR system adopts spectrum sharing model, the interference introduced to the PU bands should be carefully controlled in a tolerable range, generally measured by the interference temperature limit [9]. The transmission rate of the nth subchannel used by the kth SU is

$$r_{k,n} = \log\left(1 + \frac{P_{k,n}\left|h_{k,n}\right|^2}{\Gamma\left(N_0 W / N + I_{k,n}^{PS}\right)}\right),$$

where $h_{k,n}$ is the channel gain from the AP to the receiver of the kth SU. Γ is the SNR gap, related to a given BER requirement with $\Gamma = -\ln(5BER)/1.5$ for an uncoded MQAM [8]. The interference caused by the lth PU on the nth subchannel used by the kth SU is $I_{k,n,l}^{PS}$, which can be regarded as noise and measured by the receiver. For notation brevity, let $H_{k,n}$ denote the normalized SNR dividing the SNR gap,

$$H_{k,n} = \frac{\left|h_{k,n}\right|^2}{\Gamma\left(N_0 W / N + \sum_{l=1}^{L} I_{k,n,l}^{PS}\right)}.$$

The rate and transmission power of the kth SU are given by

$$R_k = \sum_{n=1}^{N} \rho_{k,n} \log\left(1 + p_{k,n} H_{k,n}\right), \quad P_k = \sum_{n=1}^{N} \rho_{k,n} p_{k,n},$$

where $\rho_{k,n}$ is the subchannel assignment index. If the nth subchannel is allocated to the kth SU, we have $\rho_{k,n} = 1$; otherwise, $\rho_{k,n} = 0$.

According to the analysis in Sect. 4.1, the circuit power, including static part and dynamic part, can be simplified into the static part in energy-efficient RA optimization, which will not affect the results. Therefore, we only consider the static part of circuit power P_s in this section and the EE in bits/Joule is

$$\eta_{EE} = \frac{\sum_{k=1}^{K} R_k}{\sum_{k=1}^{K} P_k + P_s}.$$

Here, we consider a general problem for the EE optimization in an OFDM-based multiuser CR system; that is, we try to maximize the EE of the CR system under the maximum transmission limit P_t at the CR AP. Each SU requires a minimal rate R_k^{min} to support its reliable communication, while the interference to each PU band should be always kept below its threshold I_l^{th}. Accordingly, the energy-efficient RA problem is formulated as follows,

$$\max_{pk,n,\rho k,n} \quad \eta_{EE} = \frac{\sum_{n=1}^{N} \sum_{k=1}^{K} \rho_{k,n} r_{k,n}}{\sum_{n=1}^{N} \sum_{k=1}^{K} \rho_{k,n} p_{k,n} + P_s}$$

$$s.t. \; C1 \quad \sum_{n=1}^{N} \rho_{k,n} r_{k,n} \geq R_k^{min}, \forall k$$

$$C2 \quad \sum_{k=1}^{K} \sum_{n=1}^{N} \rho_{k,n} p_{k,n} \leq P_t$$

$$C3 \quad \sum_{k=1}^{K} \sum_{n=1}^{N} \rho_{k,n} p_{k,n} I_{n,l}^{SP} \leq I_l^{th}, l = 1, \ldots, L$$

$$C4 \quad p_{k,n} \geq 0, \forall k, n$$

$$C5 \quad \rho_{k,n} \in \{0,1\}, \forall k, n$$

$$C6 \quad \sum_{k=1}^{K} \rho_{k,n} = 1, \forall n,$$

$$(4.5)$$

where $I_{n,l}^{SP}$ the interference to the lth PU by the transmission of an SU on the nth subchannel with unit power, as in Eq. (2.7). Each subchannel can only be assigned to one SU, as indicated by C5 and C6.

4.3.2   Relaxation Method

Since (4.5) formulates an intractable mixed integer programming problem, a general method to address such problems is relaxation, by which the integer variables are relaxed into continuous ones so that efficient linear/nonlinear optimization techniques can be utilized. In this case, the optimal solution to the relaxed problem is the upper bound of (4.5), because all feasible solutions to the original problem fall into the solution space of the relaxed one.

In relaxation method, we redefine $\rho_{k,n} = [0,1]$ as the fraction of the nth subchannel used by the kth SU, which implies that a subchannel is temporarily permitted to be shared by multiple SUs. Accordingly, we further introduce a variable $s_{k,n}$ to characterize the actual power consumption of the kth SU on the nth subchannel in a time frame interval. Thus, the relaxed form of the original problem can be written into

$$\max_{s_{k,n},\rho_{k,n}} \quad \eta_{EE} = \frac{\sum_{n=1}^{N}\sum_{k=1}^{K}\rho_{k,n}\log\left(1+s_{k,n}H_{k,n}/\rho_{k,n}\right)}{\sum_{n=1}^{N}\sum_{k=1}^{K}s_{k,n}+P_s}$$

$$s.t.\ C1 \quad \sum_{n=1}^{N}\rho_{k,n}\log(1+\frac{s_{k,n}H_{k,n}}{\rho_{k,n}}) \geq R_k^{min}, \forall k$$

$$C2 \quad \sum_{k=1}^{K}\sum_{n=1}^{N}s_{k,n} \leq P_t$$

$$C3 \quad \sum_{k=1}^{K}\sum_{n=1}^{N}s_{k,n}I_{n,l}^{SP} \leq I_l^{th}, l=1,\ldots,L$$

$$C4 \quad s_{k,n} \geq 0, \forall k,n$$

$$C5 \quad \rho_{k,n} \geq 0, \forall k,n$$

$$C6 \quad \sum_{k=1}^{K}\rho_{k,n} = 1, \forall n. \tag{4.6}$$

Equation (4.6) is a non-linear fractional programming problem which is still difficult to solve. Nevertheless, the relaxed form of η_{EE} has a special structure that its denominator is jointly concave in $\{s_{k,n},\rho_{k,n}\}$'s and the numerator is linear. Thus, instead addressing (4.6) directly, we turn to the equivalent transformation of its hypograph form [10]

$$\max_{s_{k,n},\rho_{k,n},y} \quad y$$

$$s.t. \quad \eta_{EE} \geq y$$

$$C1 \sim C6 \quad \text{in (4.6)}$$

$$y \geq 0, \tag{4.7}$$

where the inequality $y \geq 0$ is determined by the domain $\eta_{EE} \geq 0$. Such transform guarantees the equivalence between (4.6) and (4.7). Indeed, (4.7) can be geometrically described as an optimization problem in the "graph space" of $\{s_{k,n},\rho_{k,n},y\}$ with respect to the problem (4.6). In other words, we maximize the variable y over the hypograph of η_{EE} under the constraints in (4.6), which is equivalent to maximize η_{EE} in (4.6) directly. Consequently, the optimal solutions $s_{k,n}^*$'s and $\rho_{k,n}^*$'s to the problem (4.7) are also the optimal solutions to (4.6).

Let $x = \{s_{1,1},\rho_{1,1},s_{1,2},\ldots,\rho_{K,N},y\}$ denote the optimization variable of problem (4.7), the inequality constraint $\eta_{EE} \geq y$ can be converted into a convex form

$$\eta_{EE}(s_{k,n},\rho_{k,n}) \geq y \Leftrightarrow \varphi(x) \geq 0$$

with $\varphi(x) = \sum\limits_{n=1}^{N}\sum\limits_{k=1}^{K} \rho_{k,n} \log(1 + \dfrac{s_{k,n}H_{k,n}}{\rho_{k,n}}) - y(\sum\limits_{n=1}^{N}\sum\limits_{k=1}^{K} s_{k,n} + P_s)$. Then, the problem can be further converted into a convex form,

$$\max_{s_{k,n},\rho_{k,n},y} \quad y$$

$$s.t. \qquad \varphi(s,\rho,y) \geq 0$$
$$C1 \sim C6 \text{ in } (4.6)$$
$$y \geq 0. \tag{4.8}$$

Since there are fully developed algorithms to tackle such kind of problems as, it becomes optimistic to work out the optimal solution to the relaxed problem by solving the above problem.

Generally, barrier method is one of the standard techniques to solve convex optimization problems. However, the complexity, mainly incurred by the computation of Newton step via matrix inversion, is roughly bounded by $O((2KN + N)^3)$ for our considered problem. Since there are always thousands of subchannels in practical OFDM systems, such a computational cost is unacceptable, especially for the RA problem that should be tackled in an online manner. To overcome this bottleneck of standard barrier method, we try to speed up the calculation of Newton step by exploiting the special structure of the problem.

To adopt barrier method, a preparatory procedure is required to transform the objective function y in into a twice differentiable function $U(y)$. To preserve the convexity of problem, $U(y)$ should be monotone increasing and concave in y. For instance, we take $U(y) = \log(1 + y)$ in this problem. Then, the complete form of the optimization problem is

$$\max_{s_{k,n},\rho_{k,n},y} U(y) = \log(1 + y)$$

$$s.t.\, C1 \quad \varphi(s,\rho,y) \geq 0$$

$$C2 \quad \sum_{n=1}^{N} \rho_{k,n} \log\left(1 + \frac{s_{k,n}H_{k,n}}{\rho_{k,n}}\right) \geq R_k^{\min}, \forall k$$

$$C3 \quad \sum_{k=1}^{K}\sum_{n=1}^{N} s_{k,n} \geq P_t$$

$$C4 \quad \sum_{k=1}^{K}\sum_{n=1}^{N} s_{k,n} I_{n,l}^{SP} \leq I_l^{th}, l = 1,\dots,L$$

$$C5 \quad s_{k,n} \geq 0, \forall k,n$$

$$C6 \quad \rho_{k,n} \geq 0, \forall k,n$$

$$C7 \quad \sum_{k=1}^{K} \rho_{k,n} = 1, \forall n,$$

$$C8 \quad y \geq 0. \tag{4.9}$$

For the above convex optimization problem, barrier method is carried out to work out its optimal solution. In barrier method, all inequality constraints are converted into a logarithmic barrier function $\phi(x)$,

$$\phi(x) = -\log \phi(x) - \log y - \sum_{i=1}^{K+L+1} \log f_i - \sum_{k=1}^{K}\sum_{n=1}^{N} (\log s_{k,n} + \log \rho_{k,n}),$$

where

$$f_i = \begin{cases} P_t - \sum_{k=1}^{K}\sum_{n=1}^{N} s_{k,n}, & i = 1 \\[2mm] \sum_{n=1}^{N} \rho_{k,n} \log\left(1 + \dfrac{s_{k,n}H_{k,n}}{\rho_{k,n}}\right) - R_k^{\min}, & i = 2,\ldots,k+1, k = 1,\ldots,K \\[2mm] I_l^{th} - \sum_{n=1}^{N}\sum_{k=1}^{K} s_{k,n} I_{n,l}^{SP}, & i = l+K+1, l = 1,\ldots,L. \end{cases}$$

Thus, the optimal solution to the problem (4.9) can be approximated by the optimal solution to the following minimization problems,

$$\min \ \psi_t(x) = -t\log(1+y) + \phi(x)$$
$$\text{s.t.} \quad Ax = 1, \tag{4.10}$$

where t is the parameter to control the accuracy of the approximation and larger value of t provides more accurate solution. The equality constraints are unified into the matrix system $Ax = 1$ with $A_{n,m} = \begin{cases} 1 & m = 2(k-1)N + 2n, \forall k,n \\ 0 & \text{otherwise} \end{cases}$. The details of barrier method can be found in [10].

   In each iteration of barrier method, Newton method is generally used to solve the minimization problem (4.10) because of its quadratic convergence property. With a given parameter t, Newton step Δx and its associated dual variable v is given by

$$\begin{bmatrix} \Delta^2 \psi_t(x) & A^T \\ A & 0_n \end{bmatrix}\begin{bmatrix} \Delta x \\ v \end{bmatrix} = \begin{bmatrix} -\nabla \psi_t(x) \\ 0_v \end{bmatrix}, \tag{4.11}$$

where $\Delta x \in \mathfrak{R}^{2KN+1} 0_n \in \mathfrak{R}^{N\times N}$, $0_n \in \mathfrak{R}^{N\times N}$ and $0_v \in \mathfrak{R}^{N}$. $\nabla^2 \psi_t(x)$ and $\nabla \psi_t(x)$ are the Hessian and the gradient of $\psi_t(x)$, respectively.

   As aforementioned, using Newton method to solve the above equation will introduce awful complexity of $O((2KN+N)^3)$, which is obviously inapplicable for practical systems.

However, we find that this drawback can be eliminated by using a fast method to work out Newton step since the Hessian of $\psi_t(x)$ has a special structure as follows,

$$
\nabla^2 \psi_t(x) = \begin{bmatrix} D_{1,1} & & & \\ & \ddots & & \\ & & D_{K,N} & \\ & & & Y \end{bmatrix} + \sum_{i=1}^{K+L+1} \frac{\nabla f_i \nabla f_i^T}{f_i^2} + \frac{\nabla \varphi \nabla \varphi^T}{\varphi^2}
$$

$$
= D + \sum_{i=1}^{K+L+2} q_i q_i^T ,
$$

with

$$
D_{k,n} = \begin{bmatrix} \dfrac{1}{s_{k,n}^2} & 0 \\ 0 & \dfrac{1}{\rho_{k,n}^2} \end{bmatrix} + \left(\dfrac{1}{\varphi} + \dfrac{1}{f_{k+1}} \right) \begin{bmatrix} \dfrac{\partial^2 f_{k+1}}{\partial s_{k,n}^2} & \dfrac{\partial^2 f_{k+1}}{\partial s_{k,n} \partial \rho_{k,n}} \\ \dfrac{\partial^2 f_{k+1}}{\partial \rho_{k,n} \partial s_{k,n}} & \dfrac{\partial^2 f_{k+1}}{\partial \rho_{k,n}^2} \end{bmatrix} ,
$$

$$
Y = t/(1+y)^2 + 1/y^2 ,
$$

and

$$
q_i = \begin{cases} \dfrac{\nabla f_i}{f_i}, & i = 1, \ldots, K+L+1 \\ \dfrac{\nabla \varphi}{\varphi}, & i = K+L+2 \end{cases}
$$

It can be proved that the Hessian of $\psi_t(x)$ is positive definite because the matrix D is positive definite and all $q_i q_i^T \geq 0$.

Based on the analysis above, the matrix in the left of (4.11), denoted by H_0, is also invertible because $\nabla^2 \psi_t(x)$ is positive definite and A is a full row rank matrix. We can rewrite H_0 into the following form,

$$
H_0 = \begin{bmatrix} D & A^T \\ A & 0_n \end{bmatrix} + \sum_{i=1}^{M} g_i g_i^T , \tag{4.12}
$$

Where $g_i = [q_i \ 0_v]^T, i = 1 \ldots, M$ with $M = K+L+2$.

Consider the structure in (4.12), based on the matrix inversion lemma [11], we develop a fast method to speed up the computation of Newton step. The analysis is elaborated by an M-step derivation as follows,

Step 1	Decompose H_0, $H_0 = H_1 + g_1 g_1^T$
	$\begin{bmatrix} \Delta x \\ v \end{bmatrix} = v_1^1 - \dfrac{g_1 v_1^1}{1 + g_1 v_2^1} v_2^1$
	$H_1 v_1^1 = H_0$ and $H_1 v_2^1 = H_1$
	Figure out the Δx by solving v_1^1 and v_2^1, instead of using (4.11)
Step 2	Decompose H_1 with $H_1 = H_2 + g_2 g_2^T$
	Similarly, the two variables in Step 1 can be updated by
	$v_i^1 = v_i^2 - \dfrac{g_2 v_i^2}{1 + g_2 v_3^2} v_3^2, i = 1, 2$, where $H_2 v_i^2 = g_{i-1}, i = 1, 2, 3$
Consider step m,	
Step m	Let $H_{m-1} = H_m + g_m g_m^T$
	Update the m variables in Step $m-1$ by
	$v_i^{m-1} = v_i^m - \dfrac{g_m^T v_i^m}{1 + g_m v_{m+1}^m} v_{m+1}^m, i = 1, \ldots, m$
	where $H_m v_i^m = g_{i-1}, i = 1, \ldots, m+1$ with $H_i = D + \displaystyle\sum_{j=i+1}^{M} g_j g_j^T$

After the Mth step, it produces $M+1$ matrix systems $H_M v_i^M = g_{i-1}, i = 1, \ldots, M+1$. According to the above procedures, the m variables $v_i^{m-1}, i = 1, \ldots, m$ in Step $m-1$ can be updated by the $m+1$ variables $v_i^m, i = 1, \ldots, m+1$ in Step m. Hence, if we figure out the $M+1$ variables $v_i^M, i = 1, \ldots, M+1$, in the step M, the Newton step and the associated dual variable in step 1 can be indirectly obtained by an M-step reverse computation. More details about the fast algorithm for Newton method can be found in [12]. The complexity of the proposed barrier method is significantly reduced to $O(M^2 KN)$, which is much lower than $O((2KN + N)^3)$ if using standard matrix inversion.

Since each subchannel can only be occupied by one SU, we have to break the tie among multiple SUs on each subchannel, that is, to determine the best mechanism for subchannel assignment. Consider the optimal solution to the relaxed problem, larger $\rho_{k,n}$ indicates the kth SU are more preferred by the nth subchannel to maximize the desire performance, which can be regarded as a metric to determine the exact assignment of subchannels. In fact, for the optimal solution to the relaxed problem, most fraction $\rho_{k,n}$'s are close to either 1 or 0 for $K \ll N$, which has been proved in [13] and also shown in numerical results of experiment. Based on this fact, it is appropriate to allocate the nth subchannel to the kth SU with the maximum $\rho_{k,n}$, that is,

$$\rho_{k,n}^* = \begin{cases} 1, & k = \operatorname{argmax} \ \rho_{k,n} \\ 0, & \text{otherwise}. \end{cases} \tag{4.13}$$

Such principle for assignment guarantees that each subchannel is strictly allocated to only one SU. In the following simulation, it is demonstrated that such a simple rounding technique can achieve good solution close to the upper bound obtained by solving the relaxed problem.

To ensure the feasibility of solutions to the original problem, the power across subchannels should be reallocated with a given subchannel allocation. After the rounding procedure for $\rho_{k,n}$'s, its corresponding solution for power allocation $p_{k,n}$'s to the relaxed problem will no longer feasible for the original problem. Without the binary variables, the problem is simplified into a non-linear fractional programming problem. Similarly, it can be converted into a standard convex optimization problem via its hypograph form, which can be efficiently addressed by the fast barrier method. The algorithm for fast power allocation is similar to the algorithm for solving the relaxed problem, which will not be elaborated in this section. Readers can refer to the earlier part of the section for more details.

4.3.3 Numerical Results

A series of numerical experiments is carried out to evaluate the performance of our proposed algorithm. Consider a multiuser OFDM-based CR system with all SUs randomly located in a 3×3 km area. Each receiver uniformly distributed in the circle within 0.5 km from its transmitter. The path loss exponent is 4, the variance of shadowing effect is 10 dB and the amplitude of multipath fading is Rayleigh. We assume that each PU's bandwidth is randomly generated by uniform distribution and the maximum value is $2W/3L$. The noise power is 10^{-13} W.

In simulation, the optimal solution of the relaxation form serves as an upper bound of the original problem[1]. The results obtained by our proposal are compared to the upper bound for different settings. If no feasible point exists, we regard it as system outage.

Figure 4.6 illustrates the EE of the CR system as a function of the transmission power limit for different numbers of subchannels. There are four SUs with 20 bit/symbol minimal rate requirement individually, sharing the whole spectrum with two PUs. The static circuit power is set to 0.25 W. Two cases of different numbers of subchannels, $N = 32$ and $N = 64$, are investigated. The numbers of SUs and PUs are four and two, respectively. The minimal rate requirement of each SU is 20 bits/symbol and the interference threshold of each PU band is 5×10^{-12} W. The static circuit power is fixed to 0.25 W. There are $N = 32$ and $N = 64$ subchannels in the two cases. As can be seen in Fig. 4.6, the EE of the CR system grows quickly at the beginning because the possibility of CR system outage can be reduced as the increase of the transmission power budget. When the transmission power budget is sufficient enough, all SUs' rate requirements can be always satisfied and the

---

[1] Note that the upper bound cannot be a feasible solution because the relaxed form of the original problem ignores the integer constraints.

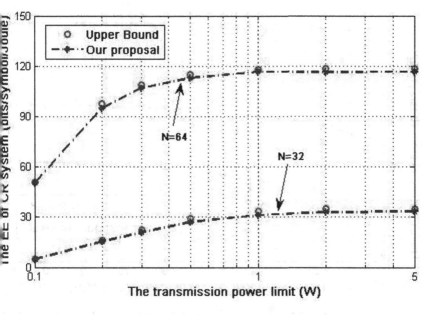

Fig. 4.6 EE versus the transmission power limit of CR system with $K = 4$, $L = 2$, $P_s = 0.25$ W and $R_k^{min} = 20$ bit/symbol

EE of the CR system keeps almost invariable. It can be explained by the EE characteristics in Fig. 4.1–4.2 that the EE can always reaches its maximum value at a certain amount of transmission power p^* since $P_t \geq p^*$. Additionally, the EE can be improved when there are more OFDM subchannels, which is a result of channel diversity in wireless environment. Notice that our proposal achieves more than 98% of the Upper Bound, indicating our proposed algorithm performs quite well for the considered problem.

We also describe the EE curve versus the interference threshold for different numbers of PUs ($L = 1$, $L = 2$ and $L = 4$) in Fig. 4.7. The number of subchannels is 64. There are four SUs with uniform rate requirement $R_k^{min} = 20$ bit/symbol. Assume that the transmission power limit at CR AP is fixed to 1 W and the static circuit power is 0.25 W, the EE of CR system increases with the increase of the interference threshold. The reason is that the lower the interference threshold, the more frequent system outage. It can be also seen from Fig. 4.7 that more PUs can diminish the EE of the CR system, which can be explained that more subchannels are interference limited in these cases and the subchannels with poor channel gains consume much power to maintain the required rates of the SUs.

From Fig. 4.6–4.7, it is evident that our proposal can always perform quite close to the upper bound for different settings. For this reason, our proposed algorithm is effective for the considered RA problem from the viewpoint of achievable EE of CR system.

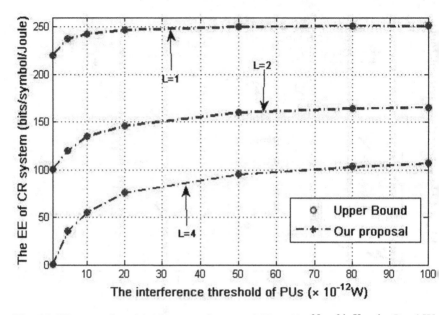

Fig. 4.7 EE versus the minimal rate requirement of SUs with $N = 64, K = 4,\ P_t = 1$ W, $P_s = 0.25$ W and $R_k^{min} = 20$ bit/symbol

Figure 4.8 illustrates the average EE of CR system as a function of the uniform rate requirements of SUs for both $K = 2$ and $K = 4$. There are 64 subchannels in CR system and two PUs in primary system. The transmission power limit is 1 W and the static circuit power is 0.25 W. For both cases, the average EE decreases with the growth of the rate requirements, since higher rate requirements not only results in exponentially increase of transmission power consumption, but also more frequent exhaustion of the radio resource, even the system outage. On the other hand, comparing the curves of the two cases, we find that the EE becomes larger with the growth of the number of SUs under relatively lower rate requirement, while more SUs can contrarily lower the EE of the CR system if the rate requirements are higher than the cut-off. Because the CR network benefits from multiuser diversity for more SUs case when the rate requirement can be easily satisfied that a subchannel is more likely to be allocated to an SU who has good channel gain over it. However, the degree of freedom for the SUs to get subchannels with good channel gains reduces in the CR system for the high rate requirements case. It means that some subchannels have to be allocated to the SUs with poor channel condition in order to meet the high rate requirement, which consumes much more power and lowers the system EE.

Convergence performance is another important indicator to evaluate the algorithms. To verify the efficiency of our proposed algorithm, we give the CDF of number of Newton iterations in the barrier method to converge for both solving the relaxed RA problem and the optimal power allocation (PA) with different settings

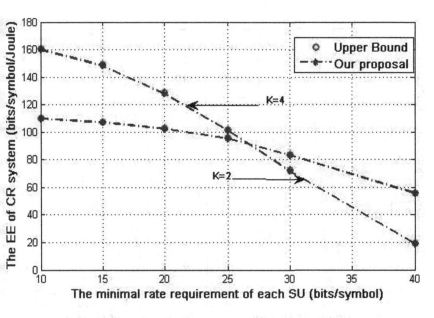

Fig. 4.8 EE versus the minimal rate requirement of SUs with $N = 64$, $L = 2$, $P_t = 1$ W and $P_s = 0.25$ W

of N in Fig. 4.9. Besides, the time cost of our proposed algorithm and the standard one which computes Newton step by matrix inversion are also investigated in Fig. 4.10. In Fig. 4.9, it is manifested that the number of Newton iterations varies in a narrow range with a given N and increases slowly when the number of subchannels gets larger. Figure 4.10 shows the average time cost (in *second*) as a function of number of subchannels over 1000 instances, for both the cases of the relaxed RA and the optimal PA. The elapsed time is counted by in-built *tic-tac* function in *Matlab*. It is remarkable that our proposed fast barrier method costs much less time for convergence, which is coincident with our analysis. Even when the number of subchannels is 256, the time cost is less than 0.2 s for the optimal PA, which can be further reduced for specialized computing platform. All these observations validate that our proposed algorithm is efficient for the considered energy-efficient RA problem.

4.4   Summary

Different from the traditional RA in OFDM-based systems, energy-efficient RA problem arises as a critical issue in green communication with focus on the utilization efficiency of energy. Actually, the increasing importance of green communication urges the operators to shift from pursuing maximum capacity to efficient energy

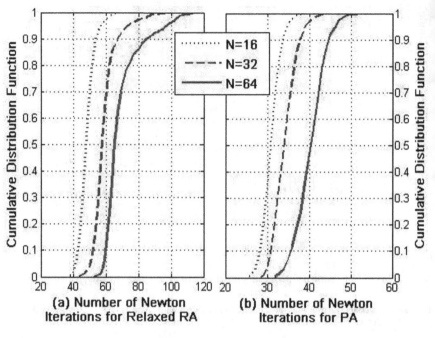

Fig. 4.9 CDF of the number of Newton iterations

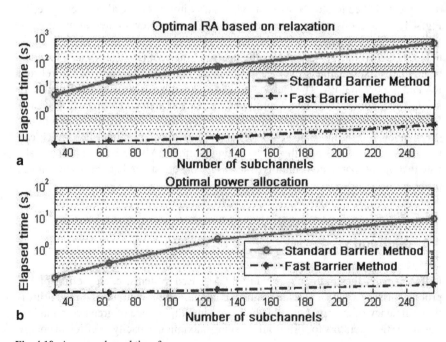

Fig. 4.10 Average elapsed time for convergence

sage for wireless networks design. Due to the limited radio resource and the strict interference limits, EE in CR system is particularly important in order to achieve the desired capacity with less expenditure and greenhouse gas emission. However, the problem becomes even more challenging for many existing algorithms are no longer suitable. In this chapter, we have discussed the basic characteristics of the EE metric and proposed several approaches to handle a general case of the problem. Furthermore, a practical energy-efficient RA problem in multiuser CR system is formulated. Based on the relaxation idea, the problem is quickly solved by the typograph-based algorithm using an improved barrier method [12, 17].

References

1. C. Han, T. Harrold, and *et al.*, "Green radio: radio techniques to enable energy-efficient wireless networks," *IEEE Commun. Mag.*, vol. 49, no. 6, pp. 46–54, June 2011.
2. D. Feng, C. Jiang, G. Lim, L. Cimini, G. Feng, and G. Li, "A survey of energy-efficient wireless communications," *IEEE Commun. Surv. & Tutor.*, vol. 15, no. 1, pp. 167–178, 2013.
3. Y. Chen, S. Zhang, S. Xu, and G. Li, "Fundamental trade-offs on green wireless networks," *IEEE Commun. Mag.*, vol. 49, no. 6, pp. 30–37, June 2011.
4. S. Sadr, A. Anpalagan, and K. Raahemifar, "Radio resource allocation algorithms for the downlink of multiuser OFDM communication systems," *IEEE Commun. Surv. & Tutor.*, vol. 1, no. 3, pp. 92–106, Sep. 2009.
5. C. Isheden and G. P. Fettweis, "Energy-efficient multi-carrier link adaptation with sum rate-dependent circuit power," in *Proc. IEEE Global Telecommun. Conf.*, Dec. 2010.
6. O. Arnold, F. Richter, G. P. Fettweis, and O. Blume, "Power consumption modeling of different base station types in heterogeneous cellular networks," in *Proc. 19th Future Network & Mobile Summit*, June 2010.
7. T. M. Cover and J. A. Thomas, *Elements of Information Theory*. New York: Wiley, 1991.
8. A. J. Goldsmith and S.-G. Chua, "Variable-rate variable-power MQAM for fading channels," *IEEE Trans. Commun.*, vol. 45, no. 10, pp. 1218–1230, Oct. 1997.
9. P. Setoodeh and S. Haykin, "Robust transmit power control for cognitive radio," *Proc. IEEE*, vol. 97, no. 5, pp. 915–939, May 2009.
10. S. Boyd and L. Vandenberghe, *Convex Optimization*. Cambridge University Press, 2004.
11. C. D. Meyer, Matrix Analysis & Applied Linear Algebra. SIAM Press, 2000.
12. S. Wang, M. Ge, W. Zhao, "Energy-Efficient Resource Allocation for OFDM-Based Cognitive Radio Networks," *IEEE Trans. Commun.*, vol. 61, no. 8, pp. 3181–3191, Aug. 2013.
13. W. Yu and J. Cioffi, "FDMA capacity of Gaussian multiple-access channels with ISI," *IEEE Trans. Commun.*, vol. 50, no. 1, pp. 102–111, Jan. 2002.
14. S. Wang, "Efficient resource allocation algorithm for cognitive OFDM systems," *IEEE Commun. Lett.*, vol. 14, no. 8, pp. 725–727, Aug. 2010.
15. Y, Wang, W, Xu, K. Yang, J. Lin, "Optimal Energy-Efficient Power Allocation for OFDM-Based Cognitive Radio Networks," *IEEE Commun. Lett.*, vol. 16, no. 9, pp. 1420–1423, Sep. 2012.
16. W. Dinkelbach, "On nonlinear fractional programming," *Management Science*, vol. 13, no. 7, pp. 492–498, 1967.
17. W. Shi and S. Wang, "Energy-Efficient Resource Allocation in Cognitive Radio Systems, "in *Proc. IEEE WCNC'13*, pp. 4531–4536, Apr. 2013.

Chapter 5
Trade-Off Between Spectral- and Energy-Efficiency

Spectral-efficiency (SE) and energy-efficiency (EE) serve as two vital metrics in dynamic resource allocation, which have been investigated separately in the previous chapters. Actually, SE and EE, however, do not always coincide and even conflict with each other sometimes [1–3]. Such characteristic makes it impossible to achieve the optimal SE and EE simultaneously all the time. Hence, how to balance the SE and EE according to the preference of the network operator is well worth studying. In light of the fact that channel capacity scales linearly with the available bandwidth but increases logarithmically with the transmission power, it is possible to trade spectral for energy efficiency, that is, to realize energy saving while guaranteeing the desired quality of service (QoS) [4].

5.1 SE-EE Relationship

In order to explore the inherent relationship between the SE and the EE, we consider the single-SU case of a CR system. There are L PUs in the primary system with total bandwidth of W. The CR system is operated with the spectrum sharing manner using OFDM modulation that the bandwidth W is divided into N subchannels in the CR system.

Assume the channel gain of the nth subchannel in the CR system is denoted by h_n, the achievable rate over the nth subchannel with transmission power p_n is

$$r_n = \frac{W}{N}\log_2\left(1 + \frac{p_n|h_n|^2}{\Gamma\left(N_0 W/N + I_n^{PS}\right)}\right),$$

where N_0 is the PSD of AWGN and Γ is the SNR gap, that is, $\Gamma = -\dfrac{In(5BER)}{1.5}$ for an uncoded MQAM with a specified BER. I_n^{PS} is the interference generated by the PU signals over the nth subchannel.

© The Author(s) 2014   93
S. Wang, *Cognitive Radio Networks,* SpringerBriefs in Computer Science,
DOI 10.1007/978-3-319-08936-2_5

Recall the definition of EE and SE,

$$\eta_{SE} = \frac{\sum_{n=1}^{N} r_n}{W} = \frac{R}{W} \qquad \eta_{EE} = \frac{\sum_{n=1}^{N} r_n}{\sum_{n=1}^{N} p_n + P_c} = \frac{R}{P + P_c}$$

where the sum rate is denoted by R and the total transmission power is P. P_c is the circuit power modeled as a linear function of system throughput with a static part P_s and a parameter ζ denoting the dynamic power consumption of unit data rate

$$P_c = P_s + \zeta R.$$

The interference constraints can be expressed as

$$\sum_{n=1}^{N} p_n I_{n,l}^{SP} \leq I_l^{th}, l = 1,...,L,$$

where $I_{n,l}^{SP}$ is the power gain from the CR transmitter to the receiver of the lth PU, which can be calculated as in [9]. I_l^{th} is the interference threshold of the lth PU. The CR system is also power-limited with the maximum transmission budget of P_t. Accordingly, the spectral or energy efficient power allocation problem can be written as the following forms, respectively,

$$\max_{p_n} \quad \eta_{SE}$$

$$s.t. \quad \sum_{n=1}^{N} p_n \leq P_t,$$

$$\sum_{n=1}^{N} p_n I_{n,l}^{SP} \leq I_l^{th}, l = 1,...,L$$

$$p_n \geq 0, \forall n \qquad\qquad (5.1)$$

$$\max_{p_n} \quad \eta_{EE}$$

$$s.t. \quad \sum_{n=1}^{N} p_n \leq P_t,$$

$$\sum_{n=1}^{N} p_n I_{n,l}^{SP} \leq I_l^{th}, l = 1,...,L$$

$$p_n \geq 0, \forall n \qquad\qquad (5.2)$$

The algorithms to solve the above problems have been elaborated in the previous chapters. However, the objectives of the SE and the EE do not always coincide with each other; that is, sometimes, it is impossible to maximize the EE and the SE simultaneously.

Theorem 5.1 For a given value of η_{SE} achieved with the power allocation $\boldsymbol{P}$, the maximum $\eta_{EE}^*(\eta_{SE}) = \max_{p_n} \eta_{EE}(\eta_{SE})$ is strictly quasiconcave in η_{SE}. Moreover, without considering the power and interference constraints, the EE, $\eta_{EE}^*(\eta_{SE})$, is first strictly increases and then strictly decreases with η_{SE}, which is maximized at $\eta_{SE} = R_{EE,max}/W$ with $R_{EE,max}$ denoting the achieved sum rate by maximizing the system EE.

Proof Without the transmission power and interference limits, denote $\mathbf{R}_1^*$, $\mathbf{R}_2^*$ and $\mathbf{R}_3^*$ as the optimal rate vectors corresponding to the overall throughput R_1, R_2 and R_3, respectively. Assume that $R_1 < R_2 < R_3$, we set,

$$\mathbf{R}_2 = \frac{R_3 - R_2}{R_3 - R_1}\mathbf{R}_1^* + \frac{R_2 - R_1}{R_3 - R_1}\mathbf{R}_3^*$$
$$= \gamma\mathbf{R}_1^* + (1 - \gamma)\mathbf{R}_3^* \tag{5.3}$$

where $\gamma = \frac{R_3 - R_2}{R_3 - R_1}$ and $0 < \gamma < 1$. The corresponding sum rate of the rate vector $\mathbf{R}_2$ is represented by R_2. According to [6–8], $P^*(\mathbf{R})$ and $I(\mathbf{R})$ is strictly convex in $\mathbf{R}$, where $P^*(\mathbf{R})$ is the optimal power corresponding to the optimal rate vector $\mathbf{R}$ if there are a sufficiently large number of subchannels.

Thus, $P^*(\mathbf{R}_2) < \gamma P^*(\mathbf{R}_1^*) + (1 - \gamma)P^*(\mathbf{R}_3^*)$. Since $\mathbf{R}_2^*$ is the optimal rate vector corresponding to the overall throughput R_2, we have $P^*(\mathbf{R}_2^*) \leq P^*(\mathbf{R}_2)$. So, we have $P^*(\mathbf{R}_2^*) < \gamma P^*(\mathbf{R}_1^*) + (1 - \gamma)P^*(\mathbf{R}_3^*)$. Thus, for any given $R(= B\eta_{SE})$, the minimum transmit power $P^*(R) = P^*(\mathbf{R}^*)$ is strictly convex in R (or η_{SE}).

Denote the superlevel set of $\eta_{EE}^*(\eta_{SE})$ as $\mathbf{S}_\beta = \left\{R \middle| \eta_{EE}^*(\eta_{SE}) \geq \beta\right\}$. $\mathbf{S}_\beta$ is equivalent to $\left\{R \middle| \beta P^*(\eta_{SE}) + \beta(P_s + \zeta R) - B\eta_{SE} \leq 0\right\}$. As a result of the convexity of $P^*(\eta_{SE})$ proved above, $\mathbf{S}_\beta$ is strictly convex in η_{SE}. Thus, $\eta_{EE}^*(\eta_{SE})$ is strictly quasiconcave and has a unique global maximum.

It is obvious that

$$\lim_{\eta_{SE}\to\infty} \eta_{EE}^m(\eta_{SE}) = \lim_{\eta_{SE}\to\infty} \max_{\eta_{SE}} \frac{B\eta_{SE}}{P^*(\eta_{SE}) + P_c}$$
$$= \lim_{P^*(\eta_{SE})\to\infty} \frac{o\left(P^*(\eta_{SE})\right)}{P^*(\eta_{SE})} = 0. \tag{5.4}$$

Thus starting from $\eta_{SE} = 0$, $\eta_{EE}^m(\eta_{SE})$ either strictly increases with η_{SE} if $\frac{d\eta_{EE}^m(\eta_{SE})}{d\eta_{SE}}\Big|\eta_{SE} = \eta_{SE}^{max} \geq 0$ or first strictly increases and then strictly decreases with η_{SE} if $\frac{d\eta_{EE}^m(\eta_{SE})}{d\eta_{SE}}\Big|\eta_{SE} = \eta_{SE}^{max} < 0$. The maximum EE in the SE region $\left[0, \eta_{SE}^{max}\right]$ is straightforward as indicated in Theorem 5.1.

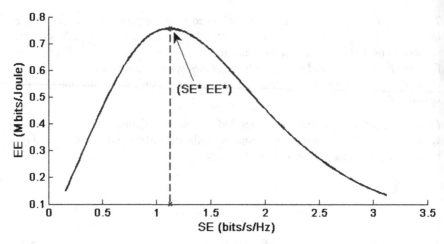

Fig. 5.1 Interrelationship between EE and SE

Actually, the EE-SE curve can be drawn by solving the following problem to show its properties as proved in ***Theorem 5.1***, that is

$$\max_{p_n} \eta_{EE}$$

$$s.t. \ \eta_{SE} = C.$$

If we enumerate all possible C and obtain their corresponding optimal solutions to the above problem, then we can figure out the EE-SE curve and the properties therein. Alternatively, the above problem can be equivalently simplified into

$$\min_{p_n} \sum_{n=1}^{N} p_n$$

$$s.t. \ \sum_{n=1}^{N} r_n = CW.$$

This problem is strictly convex in r_n's and can be solved by standard techniques, which will not be detailed here to keep this brief reasonable concise. According to the results, the curve of EE-SE is illustrated in Fig. 5.1, which is coincided to the analysis in ***Theorem 5.1***.

If imposing the transmission power limit and interference constraints on the problem, the system SE will be bounded by a maximum value η_{SE}^{max} [5], which results in two different cases of the EE-SE curves. Let η_{SE}^{*} denote the SE achieved at the optimal EE η_{EE}^{*}, if $\eta_{SE}^{max} \leq \eta_{SE}^{*}$, the maximum EE η_{EE}^{max} is achieved at the point η_{SE}^{max} as shown in Fig. 5.2 (Case 1). For this case, there is no inherent trade-off between EE and SE that both η_{SE} and η_{EE} can be maximized at the same time, since η_{EE} is monotone increasing with η_{SE} in the feasible regime. For Case 2 in Fig. 5.2, where $\eta_{SE}^{max} > \eta_{SE}^{*}$, the EE is maximized at η_{SE}^{*} while the SE is maximized at η_{SE}^{max}. In the shaded region in Fig. 5.3, it is possible to enhance the SE by reducing the EE and vice versa, which implies the potential trade-off between SE and EE.

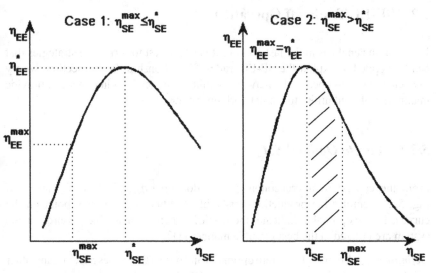

Fig. 5.2 Different scenarios of EE-SE curves

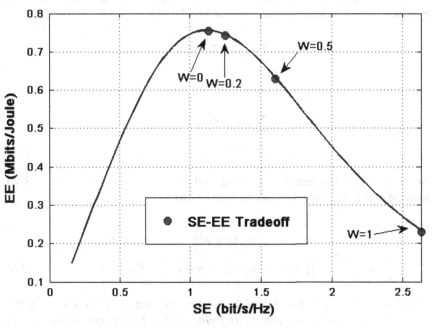

Fig. 5.3 The EE-SE relationship and the trade-off operation between SE and EE

5.2   SE-EE Trade-off Operation

Based on the analysis in the last section, it is of interest to formulate strategies that can be applied to exploit the SE-EE trade-off relationship. In this section, we try to decide where to operate the network within the tradeoff regime that can provide maximal preferred utility to the network operator.

5.2.1   Pareto Optimal Set

From the context of microeconomics, the domain $\left[\eta_{SE}^{*}, \eta_{SE}^{\max}\right]$ in the case 2 of Fig. 5.2 resembles the production possibility frontier, whereby any point on the curve is Pareto optimal. It indicates the scenario where the available resources in the system are utilized in the best possible manner [11].

Definition 5.1 A point, p is Pareto optimal if and only if there does not exist any other point p' that $\eta_{EE}(p') \geq \eta_{EE}(p)$, $\eta_{SE}(p'), \geq \eta_{SE}(p)$, and at least one ($\eta_{EE}$ or η_{SE}) has been improved.

Definition 5.2 The Pareto optimal set is the set of all Pareto optimal points.

Mathematically, we aim to maximize the EE and SE simultaneously, expressed by a multi-objective optimization problem as follows

$$\max_{p}\ \left\{\eta_{SE}(p), \eta_{EE}(p)\right\}$$

$$s.t.\ \sum_{n=1}^{N} p_{n} \leq P_{t}$$

$$\sum_{n=1}^{N} p_{n} I_{n,l}^{SP} \leq I_{l}^{th}, l = 1, ..., L$$

$$p_{n} \geq 0, \forall n. \tag{5.5}$$

According to the **Theorem 5.1** and the analysis in the previous section, the optimal solution to the above problem can be described as

$$p^{*} = \begin{cases} \operatorname{argmax}_{p}\left\{\eta_{SE}\middle|p \in p^{0}\right\} & \eta_{SE}^{*} \geq \eta_{SE}^{\max} \\ \left\{p\middle|\eta_{SE} \in \left[\eta_{SE}^{*}, \eta_{SE}^{\max}\right], p \in p^{0}\right\}, & \eta_{SE}^{*} < \eta_{SE}^{\max} \end{cases} \tag{5.6}$$

where p^{0} is the feasible region of the power allocation variables limited by the constraints in (5.5). In line with the **Definition 5.1–5.2**, all points for the case of $\eta_{SE}^{*} < \eta_{SE}^{\max}$ in (5.6) is Pareto optimal, implying the possibility of forgoing EE to gain an improvement in SE and vice versa.

5.2.2   Analysis of Utility Function

In terms of system preference, the network operator may be more inclined to either of the optimization metrics (SE or EE) to meet a certain requirement. For example, the demand to save the system expenditure indicates the first priority of EE, while the high SE is preferred than EE in a spectrum-limited case.

To facilitate system design, we should try to find a unique global solution from the Pareto optimal set. The degree of utility experienced by the network operator and a certain point of (η_{SE}, η_{EE}) can be specified by a utility function, also known as the Scalarization method for solving multi-objective optimization [12]. Scalarization method is efficient to distinguish a unique point in the Pareto optimal set. We can transform the multi-object optimization problem (5.5) into a single-object optimization problem by scalarization methods.

Define a new EE and SE trade-off metric

$$U(\mathbf{p}) = \left[\eta_{SE}(\mathbf{p})\right]^{\omega} \times \left[\eta_{EE}(\mathbf{p})\right]^{1-\omega}, \tag{5.7}$$

where ω is the preference factor with values in the range of $[0,1]$, which is set by the network operator according to the importance attached on the SE or EE. In this case, the more closer to 1 of the parameter ω, the more inclination toward to maximize the SE and vice versa for improving EE. Then, the multi-objective optimization problem can be transformed to the following single-objective problem,

$$\max_{p_n} \quad U(p)$$

$$s.t. \quad C1: p_n \geq 0, \forall n$$

$$C2: \sum_{n=1}^{N} p_n \leq P_t,$$

$$C3: \sum_{n=1}^{N} p_n I_{n,l}^{SP} \leq I_l^{th}, l = 1,...,L. \tag{5.8}$$

Notice that the case 1 in Fig. 5.2, where the optimal solution to EE and SE optimization is unique, is also included in this new-formed problem, since the utility function is also maximized at the point $\mathbf{p}^* = \text{argmax}_p \left\{ \eta_{SE} \middle| \mathbf{p} \in \mathbf{p}^0 \right\}$ when $\eta_{SE}^* \geq \eta_{SE}^{\max}$.

5.2.3   D.C. Programming

Here, we develop an efficient method to solve the optimization problem (5.8). Since the optimization objective is an exponent utility function, which is relatively difficult to deal with, we first introduce an equivalent transformation,

$$V(p) = \log U(p)$$
$$= \omega \log \eta_{SE}(p) + (1-\omega)\log \eta_{EE}(p)$$
$$= \log R - (1-\omega)\log(P + P_s + \zeta R) - \omega \log W, \tag{5.9}$$

where $P = \sum_{n=1}^{N} p_n$ and $R = \sum_{n=1}^{N} r_n$. Therefore, (5.8) is equivalent to the following problem,

$$\max_{p_n} \quad f(p) - g(p)$$
$$s.t. \ C1: p_n \geq 0, \forall n$$
$$C2: \sum_{n=1}^{N} p_n \leq P_t,$$
$$C3: \sum_{n=1}^{N} p_n I_{n,l}^{SP} \leq I_l^{th}, l = 1, ..., L \tag{5.10}$$

where $f(p) = \log R$ and $g(p) = (1-\omega)\log(P + P_s + \zeta R)$. The objective $f(p) - g(p)$ is a D.C. (difference of two convex functions) function as both $f(p)$ and $g(p)$ are concave. The gradient of $g(p)$ is $\nabla g(p) = \left(\dfrac{\partial g}{\partial p_1}, \dfrac{\partial g}{\partial p_2}, ..., \dfrac{\partial g}{\partial p_n} \right)$ where

$$\frac{\partial g}{\partial p_n} = \frac{1-\omega}{\displaystyle\sum_{n=1}^{N} p_n + P_c} \left(1 + \zeta \frac{h_n}{1 + p_n h_n} \right). \tag{5.11}$$

We propose a Frank-and-Wold (FW) procedure [13–15] which generates a sequence of improved feasible solutions. Initialized from a feasible $p^{(0)}$, $p^{(t+1)}$ at the tth iteration is generated as the optimal solution of the following convex optimization problem,

$$\max_{p} \quad f(p) - g(p^{(t)}) - < \nabla g(p^{(t)}), p - p^{(t)} >$$
$$s.t. \quad p_n \geq 0, \forall n$$
$$\sum_{n=1}^{N} p_n \leq P_t,$$
$$\sum_{n=1}^{N} p_n I_{n,l}^{SP} \leq I_l^{th}, l = 1, ..., L, \tag{5.12}$$

where $< x, y > = x^T y$. Function $g(p)$ is slowly sensitive to a change in the variable p, so $g(p)$ is well approximated by its first order approximation $g(p^{(t)}) + < \nabla g(p^{(t)}), p - p^{(t)} >$ at a fairly large neighborhood of $p^{(t)}$. Thus the nonconvex optimization problem (5.10) is well approximated by the convex optimization problem (5.12).

As function $g(p)$ is concave, its gradient $g(p^{(t)})$ is also its super-gradient, we have

$$g(P) \le g(p^{(t)}) + <\nabla g(p^{(t)}), p - p^{(t)}>. \qquad (5.13)$$

Thus the convex optimization problem (5.12) provides a well approximated lower bound maximization for the nonconvex optimization problem (5.10).Besides, as

$$f(p^{(t+1)}) - g(p^{(t+1)}) \ge f(p^{(t)}) - \left(g(p^{(t)}) + <\nabla g(p^{(t)}), p^{(t+1)} - p^{(t)}>\right)$$
$$\ge f(p^{(t)}) - g(p^{(t)}), \qquad (5.14)$$

the next solution $p^{(t+1)}$ is always better than the previous solution $p^{(t)}$.

By Cauchy theorem, since the constraint set is compact, the sequence of improved solutions $\{p^{(t)}\}$ always converges. The iterative process terminates after finite iterations at either $\left|p^{(t)} - p^{(t-1)}\right| \le \epsilon$ or $\left|U\left(p^{(t)}\right) - U\left(p^{(t-1)}\right)\right| \le \epsilon$ with threshold ϵ.

The FW procedure to solve (5.10) is summarized as follows.

Algorithm 5-1: Initialization: Set $t = 0$, choose $p^{(0)}$ and calculate $U(p^{(0)})$;

The tth iteration: Solve the convex optimization problem (5.12) to obtain the solution P^* and set $t = t+1$, $p^{(t)} = p^*$ and calculate $U(p^{(t)})$;

Stop if $\left|V(p^{(t)}) - V(p^{(t-1)})\right| \le \epsilon$.

During each iteration, the optimal solution to the standard convex optimization problem (5.12) can be worked out by standard techniques [10]. Actually, the computational efficiency can be further improved by exploiting its special structure and applying the fast barrier method, which will not be elaborated for the lack of space. Instead, the idea and procedure of fast barrier method can be found in Chap. 3 or reference [5].

5.2.4 Numerical Results

The potential trade-off between EE and SE and the effectiveness of the proposed utility function is demonstrated via a series of experiments. For simulation, we assume the whole bandwidth is 0.96 MHz, which is divided into 64 OFDM subchannels in the CR system. The path loss exponent is 4, the variance of shadowing effect is 10 dB and the amplitude of the multipath fading is Rayleigh. The PU band is randomly generated by uniform distribution with the maximum value of $2W/3L$. The noise power is 10^{-13} W and the interference threshold is 5×10^{-10} W. The static circuit power is set to 1 W and the parameter ζ in the dynamic part is fixed to 0.2 W/Mbps.

Assume the transmission power limit at the transmitter of CR system is 5 W, Fig. 5.3 illustrates the curve of EE-SE relationship and the possible trade-off operation with regard to different parameters of ω in the utility function. In line with the analysis in the context, when $\omega = 0$, efforts are made to maximize the EE, while SE is given the priority for the case of $\omega = 1$. Indeed, different values of the parameter ω imply the different degree of inclination toward to SE or EE of the system, as the

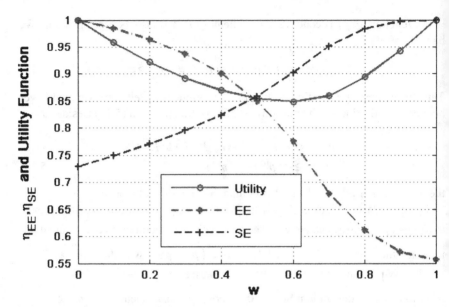

Fig. 5.4 EE, SE and the Utility under the optimal power allocation with different ω

points stamped on the EE-SE curve in Fig. 5.3. It means that the trade-off operation can be dynamically adjusted according to the preference of the network operator, as well as the instantaneous environment.

In Fig. 5.4, the normalized EE, SE and proposed utility metric achieved with the optimal power allocation is shown as a function of the parameter ω. Here, the η_{EE}, η_{SE} and $U(\eta_{EE}, \eta_{SE})$ is normalized by dividing their maximum value in order to make them dimensionless and comparable. It can be found that the normalized SE is non-decreasing and the normalized EE is non-increasing with the increase of ω, which is coincided to the results in Fig. 5.3. The relationship between SE, EE and the utility metric are also involved in Fig. 5.3, which may simplify the selection of trade-off operation.

In short, the proposed utility metric allows easier operation on a certain trade-off between SE and EE. For example, we can choose $\omega = 0.5$ for equally preferring for SE and EE. The designer or operator can take the parameter ω as the guideline for wireless communication system design.

5.3 Summary and Discussions

From the perspective of the network operators, different inclination and circumstances requires flexible strategies for system optimization. In other words, unitary pattern of EE or SE optimization, sometimes, fails to accommodate to the dynamic requirement of wireless communications. On the other hand, the interrelationship between SE and EE implies the potential trade-off between maximizing SE and EE. In this chapter we have mainly probed into the EE-SE relationship and developed the appropriate approach to exploit their possible trade-off. Specifically, inspired by the techniques in microeconomic theory, we have proposed a scalarization method to achieve the spectral- and energy-efficient trade-off operation, based on the special mathematical properties of EE-SE relationship. The utility metric, regarded as a dynamic optimization criterion for system design, can flexibly adapt to the demands of network operator and achieve a desired balance in the trade-off between SE and EE. Although we have only considered the simplified single-SU case, the proposed methods for trade-off operation can be further extended to more complicated networks, such as the multiuser OFDM systems.

In this brief, we discussed the resource allocation problems in OFDM-based cognitive radio system [16–20]. Spectral efficiency and energy efficiency are investigated extensively, as well as the tradeoffs between them. By exploiting the structure of the formulated problems, we developed a series of efficient algorithms to deal with the intractable optimization tasks, which can yield remarkable capacity gains with reasonable complexity for practical wireless networks. Since cognitive radio techniques show great potentials in many unfolding fields, such as beyond 4G cellular network and smart grid, we believe our research results shed some insights on how to design spectral- and energy-efficient cognitive radio systems in these fields, which can be interesting and promising research directions in future work.

References

1. Y. Chen, S. Zhang, S. Xu, and G. Y. Li, "Fundamental tradeoffs on green wireless networks," *IEEE Commun. Mag.*, vol. 49, no. 6, pp. 30–37, June 2011.
2. F. Meshkati, H. V. Poor, and S. C. Schwartz, "Energy-efficient resource allocation in wireless networks," *IEEE Signal Process. Mag.*, vol. 24, no. 3, pp. 58–68, May 2007.
3. G. Miao, N. Himayat, G. Y. Li, and A. Swami, "Cross-layer optimization for energy-efficient wireless communications: a survey," *Wireless Commun. Mobile Comput.*, vol. 9, no. 4, pp. 529–542, Apr. 2009.
4. C. Han, T. Harrold, S. Armour, I. Krikidis, *et al.*, "Green radio: radio techniques to enable energy-efficient wireless networks,"*IEEE Commun. Mag.,* vol. 49, no. 6, pp. 46–54, June 2011.
5. S. Wang, F. Huang, and Z. Zhou, "Fast power allocation algorithm for cognitive radio networks," *IEEE Commun. Lett.*, vol. 15, no. 8, pp. 845–847, Aug. 2011.
6. K. Seong, M. Mohseni, and J. M. Cioffi, "Optimal resource allocation for OFDMA downlink systems," in *Proc. IEEE Int. Symp. Inf. Theory*, July 2006.
7. W. Yu and R. Lui, "Dual methods for nonconvex spectrum optimization of multicarrier systems," *IEEE Trans. Commun.*, vol. 54, no. 7, pp. 1310–1322, July 2006.
8. Z. Luo and S. Zhang, "Dynamic spectrum management: complexity and duality," *IEEE J. Sel. Topics Signal Process.*, vol. 56, no. 10, pp. 57–73, Feb. 2008.
9. S. Wang, "Efficient resource allocation algorithm for cognitive OFDM systems," *IEEE Commun. Lett.*, vol. 14, no. 8, pp. 725–727, Aug. 2010.
10. S. P. Boyd and L. Vandenberghe, *Convex Optimization*. Cambridge University Press, 2004.

11. R. S. Pindyck and D. L. Rubinfeld, *Microeconomics*. Upper Saddle, NJ: Prentice-Hall, 2008.
12. R. T. Marler and J. S. Arora, "Survey of multi-objective optimization methods for engineering," *Structural and Multidisciplinary Optimization*, vol. 26, no. 6, pp. 369–395, 2004.
13. M. Frank and P. Wolfe, "An algorithm for quadratic programming," *Naval Res. Log. Quart*, vol. 3, pp. 95–110, 1956.
14. P. Apkarian and H. D. Tuan, "Robust control via concave optimization: local and global algorithms," *IEEE Trans. Automatic Control*, vol. 45, pp. 299–305, Feb. 2000.
15. P. Apkarian and H. D. Tuan, "Concave programming in control theory," *J. Global Optimization*, vol. 15, no. 4, pp. 343–370, 1999.
16. M. Ge and S. Wang, "Fast optimal resource allocation is possible formultiuser OFDM-based cognitive radio networks with heterogeneousservices," *IEEE Trans. Wireless Commun.*, vol. 11, no. 4, pp. 1500–1509, Apr. 2012.
17. S. Wang, Z.-H. Zhou, M. Ge and C. Wang, "Resource allocation for heterogeneous cognitive radio networkswith imperfect spectrum sensing," *IEEE J. Sel. Areas Commun.*, vol. 31,no. 3, pp. 464–475, 2013.
18. S. Wang, M. Ge and W. Zhao, "Energy-Efficient Resource Allocation for OFDM-based Cognitive Radio Networks," *IEEE Trans. Commun.*, vol. 61, no. 8, pp. 3181–3191, Aug. 2013.
19. S. Wang, M. Ge, C. Wang, "Efficient Resource Allocation for Cognitive Radio Networks with Cooperative Relays," *IEEE J. Sel. Areas Commun.*, vol. 31, no. 11, pp. 2432–2441, Nov. 2013.
20. S. Wang, Z.-H. Zhou, M. Ge and C. Wang, "Resource Allocation for Heterogeneous Multiuser OFDM-based Cognitive Radio Networks with Imperfect Spectrum Sensing," In *Proc. IEEE INFOCOM'12*, pp. 2264–2272, Mar. 2012.